AQUACULTURE WITH ARTIFICIAL INTELLIGENCE

AQUACULTURE WITH ARTIFICIAL INTELLIGENCE

Emerging Technologies for Sustainable Industry Development

BASKARALINGAM VASEEHARAN
Alagappa University, Karaikudi, India

ASHOKKUMAR SIBIYA
Thiagarajar College, Madurai, India

PALMY JESUDHASAN
USDA Agriculture Research Service, USA

CHANDRABOSE SELVARAJ
Amet University, Chennai, India

ELSEVIER

ACADEMIC PRESS
An imprint of Elsevier

Academic Press is an imprint of Elsevier
125 London Wall, London EC2Y 5AS, United Kingdom
50 Hampshire Street, 5th Floor, Cambridge, MA 02139, United States

ISBN: 978-0-443-33704-8

For information on all Academic Press publications visit our website at https://www.elsevier.com/books-and-journals

Publisher: Mara Conner
Acquisitions Editor: Vaishali Singh
Editorial Project Manager: Kritika Sharma
Production Project Manager: Fizza Fathima
Cover Designer: Bakyalakshmi S

Typeset by TNQ Tech

Contents

Foreword I

I am pleased and honored to write the foreword for *Aquaculture with Artificial Intelligence: Emerging Technologies for Sustainable Industry Development*. This book arrives at a critical time, as aquaculture faces growing pressure to increase production while ensuring sustainability, animal health, and environmental responsibility.

Aquaculture plays a vital role in global food security, yet it continues to face persistent challenges, including disease outbreaks, climate change, and resource constraints. In this context, artificial intelligence offers powerful and timely solutions. This book brings together fundamental concepts and advanced applications, learly demonstrating how AI-driven tools such as disease monitoring, smart feeding, and data-driven decision-making can transform aquaculture into a more efficient and responsive system.

What sets this book apart is its ability to bridge scientific innovation with practical application. It presents interdisciplinary approaches that not only enhance productivity but also reduce reliance on antibiotics and support sustainable practices. The content is accessible and relevant, making it valuable for researchers, students, industry practitioners, and policymakers alike.

I believe this book will inspire meaningful progress in both research and practice. It stands as an important contribution toward a more resilient, sustainable, and technology-enabled future for aquaculture.

Prof. Dr. Ina Salwany Md Yasin
Department of Aquaculture
Faculty of Agriculture
and Deputy Director,
Institute of Bioscience
Universiti Putra Malaysia

Foreword II

It is with great pleasure and a deep sense of academic fulfillment that I write the foreword to this book titled ***Aquaculture with Artificial Intelligence: Emerging Technologies for Sustainable Industry Development.*** The advent of Artificial Intelligence (AI) and associated digital technologies has opened unprecedented opportunities to transform traditional aquaculture practices into highly efficient, precise, and environmentally responsible systems. It begins with a strong foundation through an introductory understanding of aquaculture and progresses into recent technological advancements shaping the field. The chapters on digitalization, smart aquaculture systems, and the introduction of AI lay a critical groundwork for readers to appreciate how data-driven approaches are revolutionizing aquaculture operations. Particularly noteworthy are the discussions on AI and machine learning algorithms tailored for aquaculture applications, offering insights into predictive analytics, automation, and intelligent decision-making. The integration of bioimaging with AI for disease detection and monitoring represents a significant leap toward ensuring aquatic animal health and minimizing losses. Similarly, the incorporation of drone technologies for automated feeding and monitoring reflects the innovative convergence of engineering and biological sciences. The book further explores interdisciplinary advancements such as AI-integrated bioengineering, Internet of Things (IoT) applications, and next-generation genome sequencing, which collectively enhance productivity, genetic improvement, and sustainability. The chapters on AI-driven drug design and emerging computational frameworks, including blockchain, cloud computing, and quantum technologies, highlight the transformative potential of digital ecosystems in ensuring transparency, traceability, and efficiency in aquaculture industries. Importantly, the volume does not limit itself to theoretical advancements but extends its focus to industrial applications and future directions. It emphasizes how AI integration supports scalable, economically viable, and environmentally sustainable aquaculture practices. The forward-looking perspectives presented here will undoubtedly inspire researchers, academicians, policymakers, and industry stakeholders to adopt innovative solutions for long-term sectoral growth. I am confident that this book will significantly contribute to advancing knowledge and fostering sustainable development in this vital field. I extend my best wishes for the success of this publication and trust that it will stimulate further research, innovation, and collaboration at the intersection of aquaculture and artificial intelligence.

Prof. P. Ramasamy
Former Vice Chancellor
Alagappa University,
Karaikudi-630003,
Tamil Nadu, India

Foreword III

It is a great honor to write this foreword for the book entitled ***Aquaculture with Artificial Intelligence: Emerging Technologies for Sustainable Industry Development***. I sincerely believe that this valuable work is the result of extensive research and the dedicated efforts of the authors who have contributed to this book. The book presents a comprehensive understanding of aquaculture, with a particular focus on the application of artificial intelligence, and highlights its importance in promoting sustainable industry growth. Each chapter focuses on a distinct area of aquaculture and modern technologies. I am confident that readers will have clear insights into both fundamental concepts and practical applications. I appreciate the authors for their commitment to collecting, organizing, and presenting relevant information in such a comprehensive manner.

Aquaculture is a rapidly expanding sector that contributes significantly to meeting global demands for nutrient-rich animal protein. With the growing need for increased production and sustainability, the implementation of advanced technologies has become essential. In this context, artificial intelligence and machine learning can play a crucial role in reshaping aquaculture practices by enabling better decision-making, accurate monitoring, and efficient resource utilization.

The book begins with the fundamentals of aquaculture and progresses into recent technological advancements, including the development of smart aquaculture systems. It clearly explains how artificial intelligence and machine learning techniques can be applied to improve aquaculture operations. Important topics such as the use of bioimaging for disease detection, drone-based feeding systems, and AI-supported bioengineering for enhancing aquatic animal health are discussed in detail. Moreover, the utilization of the Internet of Things (IoT), OMICS approaches, and AI-driven drug discovery methods demonstrates the interdisciplinary nature of modern aquaculture research. The discussion on blockchain, cloud platforms, and big data highlights the role of digital tools in improving efficiency and management within aquaculture systems. The book also addresses industrial applications and provides valuable insights into the future directions of aquaculture with artificial intelligence.

I extend my sincere appreciation to the authors for their academic contributions in covering a wide spectrum of topics related to aquaculture and artificial intelligence, including their concepts, applications, and future potential. I am confident that this book will be highly beneficial for students, researchers, aquaculture practitioners, and industry professionals and will inspire further innovation in the field of sustainable aquaculture.

J.K. Jena
Deputy Director General (Fisheries Science)
Indian Council of Agricultural Research
Krishi Anusandhan Bhavan II,
Pusa, New Delhi, India 110012

1

Aquaculture—An introductory text

Introduction

Aquaculture is simply the cultivation of aquatic organism under a controlled environment. This plays a significant role in satisfying the global needs, reducing the rate at which wild populations are caught and harvested, and allowing them to recover and maintain themselves. The importance extends far beyond food production, encompassing economic improvement, reducing scarcity, and ecological sustainability (Mansour et al., 2022). As the population increases, aquaculture stands as main support in ensuring food safety, giving benefits to millions of people all over the world. Still the quick aquaculture extension faces challenges, primarily in the disease occurrence area. Intensive farming practices and high-density aquaculture ecology create a breeding area for microbes that devastate fish and shellfish population rate. Viruses, bacteria, and parasites pose noteworthy pressures, resulting in considerable commercial losses and ecological devastation. Highlighting these tasks needs a complete understanding of disease dynamics and the application of strong preclusion methods. Inhibition in aquaculture contains multifaceted tactics. Biosecurity measures, including regulating water quality, nursing and handling stocking densities, and executing isolation methods, are essential in modifying disease risks. Proactive health supervision, with steady health valuation and inoculations, boosts and supports the resilience of the aquatic populations. Besides, the progressions in representatives and breeding technologies subsidize to the improvement of disease-resistant strains, minimizing the vulnerability of cultivated species (Subasinghe et al., 2003). At present, the aquaculture industry has observed a paradigm shift toward innovative activities that incorporate advanced technological skills. Exact aquaculture permitted by radars, artificial intelligence (AI), and information analytics permits culturists to regulate and manage ecological parameters in real time. Robotic feeding systems, recirculating

Aquaculture with Artificial Intelligence. https://doi.org/10.1016/B978-0-443-33704-8.00002-1

aquaculture systems (RASs), and closed-contaminated machineries reduce ecological effects and regulate source usage. These ideas not only improve efficacy but also subsidize the complete sustainability of aquaculture methods. Moreover, the incorporation of sustainable methods, including organic culture and integrated multitrophic aquaculture (IMTA), improves a harmonious coexistence between aquaculture and the surrounding ecological system (Khanjani et al., 2022). IMTA, for example, includes various species in immediacy to make a mutual bonding, where the byproducts of one species serve as feed for another. This minimizes waste and improves overall system resilience (Authority, 2014). In the aquaculture field, the combination of AI and machine learning (ML) has accompanied in a new era of innovative activities, altering aquatic path cultivation and maintenance. This chapter explores the complex applications of AI and ML in aquaculture, showing the thoughtful effects these technologies have on improving efficiency, sustainability, and complete output. One of the essential improvements in aquaculture activities falls in the field of forecasting analytics, where ML algorithms evaluate numerous datasets encircling ecological variables, water quality parameters, and past yield information to predict ideal situations for aquatic species. These prophetic models allow farmers to preventively address possible events, including disease outbreaks, ensuring active measures are taken to sustain the health and well-being of the aquatic organisms (Mandal & Ghosh, 2023). Moreover, AI-powered monitoring systems have evolved as crucial tools for real-time observation and regulation within aquaculture facilities. Integrated sensors continuously gather information on parameters such as temperature, oxygen levels, and feeding, giving immediate feedback to AI algorithms. With constant learning, these systems adjust to the active nature of aquaculture ecology, regulating nursing timetable, modifying stressors, and mechanizing the daily errands. Exact aquaculture is totally different in which AI and ML shine, permitting targeted and efficient resource utilization. Self-directed underwater topography and the detection of ideal areas for aquaculture systems increase the water flow, nutrient supply, and overall operational efficacy. In disease management, AI plays a key role in initial recognition and quick response. ML models investigate patterns in fish behavior, health detectors, and ecological issues to find refined symptoms of disease onset. This active method includes applying appropriate involvements, thereby minimizing the effect of diseases. Furthermore, automation in aquaculture activities has replaced labor-intensive tasks, such as feeding, sorting, and harvesting. These automatons are equipped with processed version and ML skill, permitting them to adapt to the exclusive features of each aquaculture facility and carry out tanks with precision, reducing labor workload and minimizing human error. As we continue to assess the interactions between skills and aquaculture, the potential for novelty and growth in this area is infinite (Mohale et al., 2024). This chapter covers aquaculture, its culture systems, diseases, precaution methods, and progression in aquaculture practices using AI and ML for sustainability.

Overview of aquaculture

The term aquaculture refers to the farming of aquatic organisms—including fish, crustaceans, molluscs, and aquatic plants—which has evolved as a significant factor of worldwide food production and commercial improvement (Allsopp et al., 2012). Referring to ancient Chinese, Egyptian, and Roman cultures, aquaculture has progressed meaningfully over millennia to meet the cumulative needs for seafood while reducing pressure on wild fish stocks. At present, it encompasses a diverse array of systems, including freshwater and marine operations, land-based ponds, raceways, and offshore cages (Mansour et al., 2022). Aquaculture contributes substantially to global food security by providing a significant portion of the world's seafood supply. Moreover, it gives commercial chances for seaside societies, providing adoptive work options and generating profits. Aquaculture also experiences tasks relating to ecological effects, such as habitat destruction, water contamination, and the spread of diseases. Sustainable follow-ups, technological novelties, and strict rules are essential in highlighting these concerns and confirming the long-term capability of aquaculture as a sustainable food production system (FAO, 2020). The origin of aquaculture can be outlined back 1000 years to primeval China, where carp bred in artificial ponds. Along with the time, this method spread to nearby areas, with various species being farmed in different situations. In the current era, progressions in techniques and improved thoughts of bionetworks have pushed aquaculture to new stages, permitting the culturing of broad species in different situations, includes ponds, tanks, and seaward structure. Although aquaculture faces ecological tasks, it provides resolution to overfishing. This system should be maintained properly, or otherwise it risks to the issues, such as water contamination, habit destruction, and the introduction of nonnative species (Allsopp et al., 2013). Sustainable aquaculture methods concentrate on reducing these adverse effects via responsible site selection, efficient surplus management, and the usage of environmentally friendly feeds. Integrating aquaculture with other food production systems, such as aquaponics, further enhances sustainability, transforming waste products into nutrients for different crops. As aquaculture continues to expand, it alleviates pressure on wild fish stocks, contributes to economic development, and enhances food security by providing a reliable source of protein for a growing global population.

Cultural system in aquaculture

The cultivation of various aquatic organisms under a controlled condition is called as aquaculture (Fig. 1.1). The aquaculture systems are diverse, and they vary based on the type of species being cultivated, environmental conditions, and the goals of the aquaculture operation (Khanjani et al., 2022).

Pond culture

Pond culture is the ancient and broadly applied in aquaculture systems. This adds natural or artificial ponds to farm fish, shrimp, or other aquatic organisms (Fig. 1.2). These ponds are

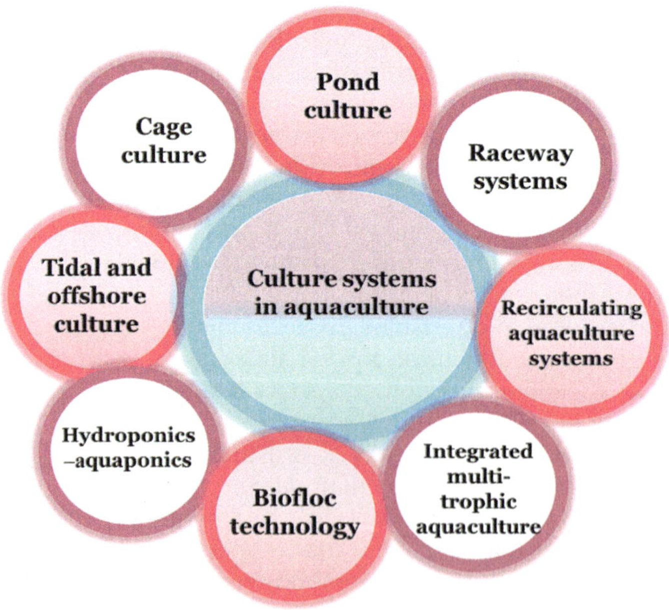

Figure 1.1 Various culture systems in aquaculture.

Figure 1.2 Pond culture system.

characteristically built up with earthen embankments, and water quality is maintained through ventilation, water exchange, and the usage of natural vegetation (Drozdz et al., 2020). This is mostly applicable for freshwater fish, including tilapia, catfish, and carp along with shrimp in brackish water.

Cage culture

Cage culture includes constraining fish or other aquatic organisms in buoyancy cages or pens within natural water bodies, such as rivers, lakes, or coastal areas (Fig. 1.3). They are characteristically made of nets or rigid things and are fastened to markers to make them stay underwater (Salin et al., 2023). The feed is given straightly to crate animals. This method is appropriate for different species, such as salmon, trout, sea bass, and sea bream.

Raceway systems

Raceway systems contain fine, flow-through networks or tanks which permits water to flow unceasingly, giving a precise ecology for fish cultivation. Water is supplied at one termination and pass through the other way that carries the waste and ventilates. These systems are frequently used for rearing trout and salmon. The capability of regulating and monitoring the fish

Figure 1.3 Cage culture system.

health is achieved by well-organized water usage and practical water quality control (Engle et al., 2020).

Recirculating aquaculture systems

RAS includes the recirculation and reuse of water in a locked system, combining the motorized and biological filtration to keep water quality (Fig. 1.4). Tanks, biofilters, aeration systems, and water treatment techniques are essential apparatus (Ahmed & Turchini, 2021). RAS permits exhaustive and high-density fish rearing. This system is suitable for various fish types, such as tilapia, catfish, salmon, and sturgeon.

Integrated multitrophic aquaculture (IMTA)

IMTA can be defined as the maintainable method that integrates the culturing of various species in the same aquatic environment to extend the source usage and minimize ecological impact. Characteristically, this includes farming of fish, shellfish, and seaweed in a combined state, where the remains from one species act as nutrients for others. So, this enrichens the complete system efficiency, minimizes the ecological effects, and improves biodiversity (Nederlof et al., 2022).

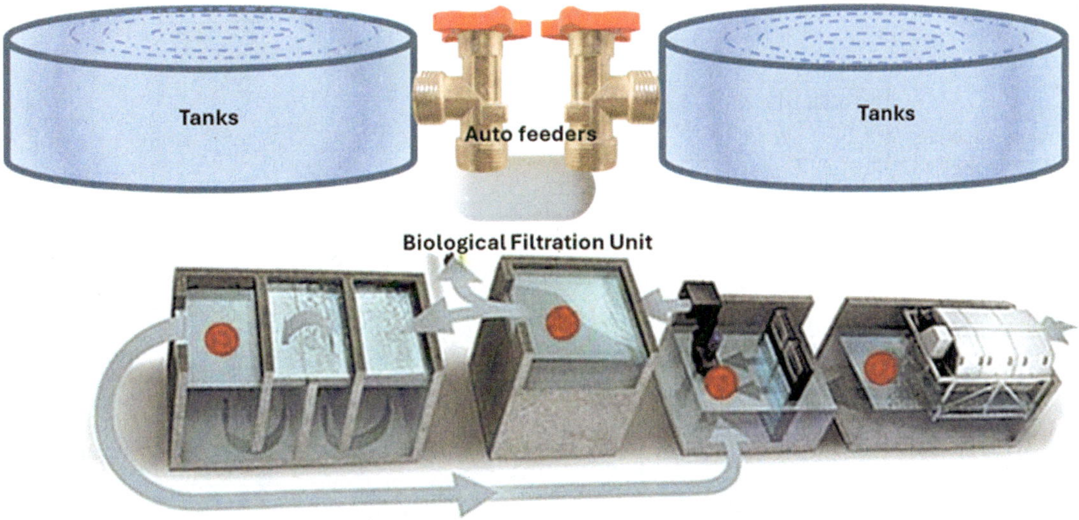

Figure 1.4 Recirculating aquaculture systems.

Biofloc technology

BFT is systematic structure that upgrades the synthesis of microbial aggregates (Biofloc) to uphold the quality of water source and gives the additional food resources for cultivated organisms. Aeration, carbon supplementation, and microbial management are crucial elements. BFT is commonly used in shrimp farming. It improves water quality, reduces reliance on external feed, and enhances nutrient recycling (Ogello et al., 2021).

Hydroponics–aquaponics system

The combination of aquaculture with hydroponics leads to the aquaponics, where the fish waste provides nutrients for plants, and the plants, in turn, help purify the water for the fish (Fig. 1.5). This setup uses tanks or ponds for the fish and hydroponics beds or containers for the plants. It fosters a symbiotic relationship between the fish and plants (Rajalakshmi et al., 2022) (Fig. 1.6), enabling the efficient use of resources, reducing environmental impact, and allowing for the simultaneous production of both fish and vegetables.

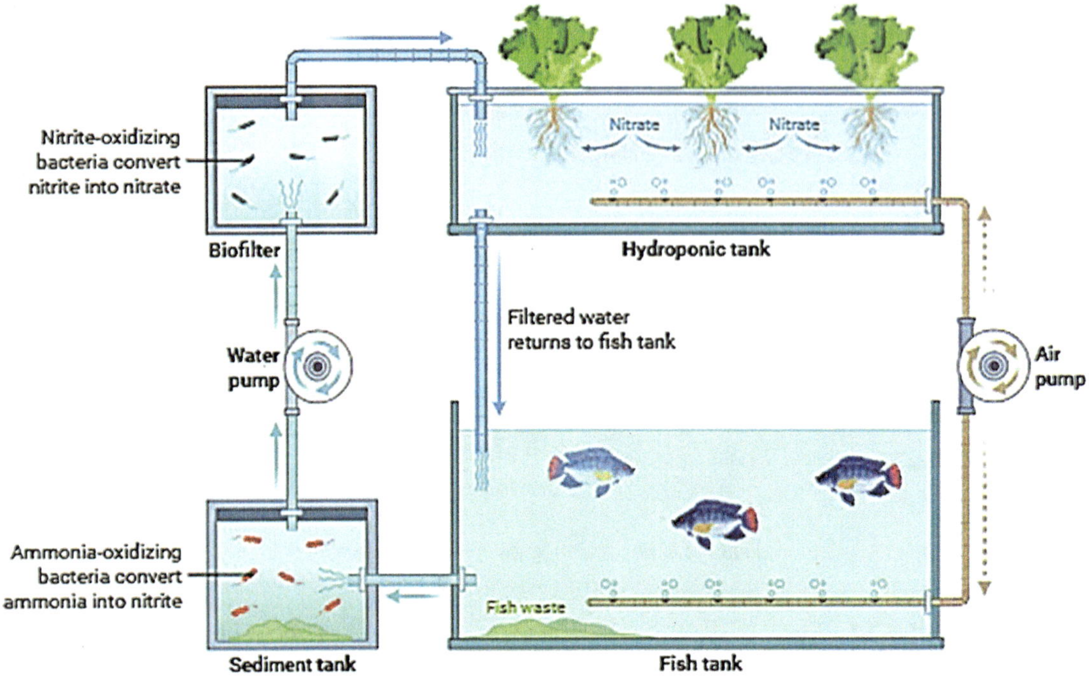

Figure 1.5 Hydroponics–aquaponics systems.

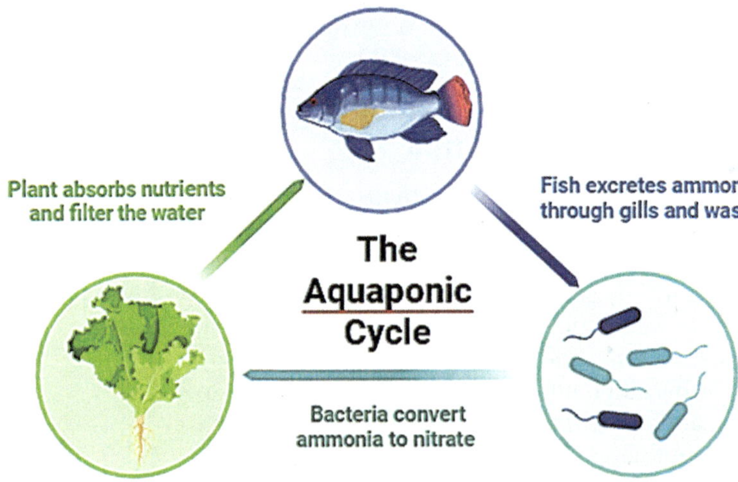

Figure 1.6 The aquaponics cycle.

Tidal and offshore aquaculture

This process involves the cultivation of marine species in open waters, capitalizing on natural tidal movements and ocean currents. Floating structures, sea cages, and mooring systems are used. It is often for high-value species, such as tuna or cobia. It uses natural water bodies, minimizes environmental impact, and allows for the cultivation of species requiring larger spaces (Morro et al., 2022). Every aquaculture system has its profits and risks, and the option of system falls on the aspects that includes target species, accessible sources, ecological deliberations, and the aims of the aquaculture activities. Sustainable applications, appropriate supervision, and constant organization are important for the long-term achievement of aquaculture systems.

Diseases and preventive measures

Disease exhibits an important menace to the aquaculture industry, and precautionary events are essential for regulating the health and sustainability of aquaculture activities. Aquaculture systems are places where all the aquatic organisms are kept in small spaces and are prone to stress and disease. To report this, aquafarmers take a chance on using the antibiotics for disease control (Subasinghe et al., 2023). Still, the regular use of antibiotics directs to the concerns about the growth and spread of antibiotic resistance (Lieke et al., 2020). To escape from these tasks, the sector needs to investigate various technological revolutions and sustainable policies. These include developing the workshop facilities, analytical skill, and regulated and healing methods to handle disease occurrences effectually. In addition to this, the usage of nanotechniques in aquaculture has projected a potential in disease supervision. This evolving area has proved progressions in water disinfection, food conversion, fish health, and management systems. These progressions display fresh chances for improving disease regulation in aquaculture and qualifying the perils related by means of antibiotic usage (Mugimba et al., 2021). Using nanotechnology in aquaculture, fresh and novel ideas have been framed to deduce diseases and encourage sustainable production. In the present world, the importance of precise disease inhibition methods in aquaculture cannot be exaggerated. In the aquaculture filed, machine-learning techniques have raised as a controlling tackle to improve disease management in this situation (Nasr-Eldahan et al., 2021). Through the integration of innovative algorithms, ML allows to study the numerous datasets, which include ecological parameters, water quality, and past disease occurrence records. They find out the outlines and relationship that may be unnoticed by traditional techniques, and give complete thoughts on the factors influencing disease outbreaks (Subasinghe et al., 2023). ML improves forecasting models that predict the possible disease risk and allows for proactive actions to be applied. In addition, real-time regulating systems, controlled by ML algorithms, constantly spot eccentricities from normal conditions, confirming quick response to upcoming threats (Gladju et al., 2022). Moreover, these techniques lead to adjusting the feed management and ecological situations, upgrading complete system resilience. Essentially, the integration of ML in aquaculture not only improves disease prevention but also boosts a more effective and sustainable method to handling the aquaculture ecosystems.

Reproduction and animal rearing

Reproduction and animal rearing are the basic factors of aquaculture, which provides the sustainable yield of fish, crustaceans, and other aquatic species. Controlled breeding of chosen species in aquaculture in the monitored area results in effective reproduction. The above method wants working on maintaining ecological conditions, including temperature, light, and water quality to trigger the reproductive behavior and regulate the existence of young ones (Azra et al., 2022). In aquaculture, numerous reproductive methods are applied, starting from natural spawning captivity to artificially inducing breeding technique with hormonal manipulation. Regulated reproduction permits farmers to breed the chosen individuals with selective characteristics that include quick growth rates, disease resistance, or enhanced fillet quality. This chosen breeding improves the overall act of the farmed species and subsidizes to the hereditary enhancement of aquaculture stocks (Arechavala-Lopez et al., 2022). After the completion of successful reproduction phase, it is followed by the animal rearing phase. Aquaculture system applies various techniques for rearing larvae, fingerlings, or juvenile organisms, depending on the species. Hatcheries are mostly involved in rearing aquatic larvae in a monitored situation, giving an ideal setting for growth and development. This allows farmers to clearly regulate the parameters, including water temperature, salinity, and nutrition to confirm the health and potential of the young organisms. In aquaculture, the evolution from the larvae to juvenile, resulting in the saleable size, is a fine procedure that needs cautious supervision. Ponds, raceways, tanks, and cages are in rearing systems that are designed to facilitate the exact requirements of the cultured species. Appropriate nutrition, disease management, and water quality maintenance are essential parts of effective animal rearing, confirming ideal growth and minimizing stress. Aquaculture's prominence on the reproduction and animal rearing attaches with the trade's assurance to sustainability (Azra et al., 2022). Resilience on wild-caught broodstock can be reduced by refined breeding techniques, and by applying efficient rearing practices, aquaculture contributes to biodiversity conservation and pressure reduces on natural population. In addition, the aquaculture facilities of controlled environment permit better disease management, thereby minimizing the risk of outbreaks that could affect both farmed and wild populations. As the aquaculture industry continuous to evolve, advancements in reproductive technologies and rearing methodologies will contribute to increased efficiency, reduced environmental impact, and the continued growth of this essential sector in global food production. In the domain of reproduction, ML algorithms help in the analysis of complex datasets related to reproductive behavior, genetics, and environmental factors affecting aquatic species (Selvaraj et al., 2021). This tells in advance the optimal breeding conditions, identifies the genetic markers associated with the desirable traits, and enhances the selective breeding programs for improved stock.

Also, ML models detect and promote environmental parameters, such as water quality and temperature, to ensure an ideal habitat for reproduction. In the context of animal rearing, ML applications aid in the automated monitoring of feeding regimes, disease detection, and growth patterns. Leveraging real-time data, these systems adjust feeding schedules, detect early signs of diseases, and optimize the overall management of aquaculture operations (Azra et al., 2022). Ultimately, the integration of ML in aquaculture enhances efficiency, productivity, and sustainability in both reproduction and animal-rearing processes, contributing to the advancement of this vital industry.

Nutrition and growth of aquatic animals

Overall health, reproductive success, and resilience to environmental stressors of aquatic animals were influenced by the nutrients, which play a pivotal role in the growth and development. In the dynamic ecosystem of aquatic environments, the availability and composition of food sources have a significant impact on the well-being of diverse aquatic species (Kandathil et al., 2020). Based on the factors such as size, age, reproductive status, and habitat, nutritional requirements vary. One of the key factors affecting the growth of the aquatic animals is the balance of essential nutrients in their diet. Critical components such as proteins, carbohydrates, fats, vitamins, and minerals must be present in the appropriate amounts to ensure optimal growth. For muscle development and metabolic functions, proteins are vital, while fats contribute as an energy source to the structure of cell membranes. Quick energy is provided by carbohydrates, and roles in physiological process are played by vitamins and minerals. The source and form of nutrition are also crucial considerations. Herbivores consume primary plant-based nutrients, while carnivores consume animal proteins. On the contrary, omnivores consume both plant-based and animal proteins. The availability of appropriate prey or vegetation in their habitat directly influences the nutritional intake of aquatic organisms. Ecological factors such as water temperature, salinity, and pH further influence the dietary needs of aquatic animals (Lall & Kaushik, 2021). The availability and digestibility of nutrients are influenced by these factors and affect the overall efficiency of nutrient utilization for growth and maintenance. The fluctuations in environmental conditions can alter the composition and abundance of prey or plant species, which directly affects the nutritional options for the aquatic organisms. Cultivation of marine animals by understanding and proving the paramount importance is paramount importance because of the controlled meet human demand (Kandathil et al., 2020). For maximizing growth rates, reproduction, and overall health in aquaculture settings, it is essential to formulate the balanced and species-specific diets. The nutritional content of feed, feed frequency, and feeding methods to meet the specific requirements of the target species must be considered by the aquaculturists. A thorough insight of the nutritional requirements, dietary selections, and sustainable aquafarm methods is essential (Dawood, 2021). Addressing these factors, experts contribute to the well-being and productivity of aquatic animals, confirming heath and strength of these vital components of aquatic ecosystems. There is an ever-improving focus on sustainability in aquaculture.

Advancements in aquaculture practices

Aquaculture, the method of farming aquatic organisms such as fish, crustaceans, molluscs, and aquatic plants, has notified important advancements in recent years. An integration of technological innovation, an increased understanding of aquatic ecosystem, and the necessary to meet the developing global demand for seafood drive these advancements (Mandal & Ghosh, 2023). There is an ever-increasing importance on sustainability in aquaculture. This includes decreasing environmental effects, including habitat destruction, pollution, and overuse of resources. Sustainable aquaculture methods intent to reduce surplus, use ecological feeds, and accept systems that reuse water and nutrients. Aquaculture is benefitting from the incorporation of different machineries, such as radars, mechanization, and data analytics. These technologies assist in tracking the water quality, automate feeding systems, optimize production processed, and manage the health parameter of aquatic organisms (Authority, 2014). Intensive production of aquatic organisms in a controlled environment with minimal water exchange is allowed by RASs. They deduct the use of larger water bodies and decreases the ecological effects by reprocessing the water and utilize by purification and treatment technologies. In the same way, IMTA systems grow numerous species together in a mutual bond, where surplus products from species provide nutrients for another. This method reduces surplus and nutrients contamination, enhances usage efficacy, and improves complete environmental health (Khanjani et al., 2022). Besides this, to improve strains of aquatic organisms with required characteristics, which include quick progress, disease resistance, and developed feed adaption proportions, chosen breeding programs are being employed. Genomic technologies, including marker-assisted and gene selection, are hastening the refining programs, ending up in the development of various other prolific and resilient species. Offshore farms are away from the coastal areas so it makes increasing interest in aquafarming. They have the potential to alleviate pressure on nearshore environments, minimize conflicts with of coastal area users, and access cleaner water with fewer disease vectors. It is said that the advancement of aquaculture practices is essential for meeting the growing demand for seafood while minimizing environmental impacts and promoting social responsibility throughout the supply chain. For further improving the sustainability and efficiency of aquaculture in the future, continued innovation and collaboration across scientific, technological, and regulatory domains will be crucial.

Introduction of AI and ML for sustainable aquaculture

AI, a part of computer science which enables machines to perform tasks typically requiring human intelligence, and ML, a subset of AI that focuses on algorithms allowing computers to learn from and make forecasting based on the data, provide a stage of applications which are applicable for improving eco-friendly aquafarming conditions. ML is highly applied in aquafarming to improve different components of aquafarming, that consist tracking, refining and decision- (Mandal & Ghosh, 2023). AI and ML technologies facilitate accurate observing and supervising of aquaculture environments. Through the comprehensive assessment of real-time data from sensors, cameras, and other Internet of Things (IoT) devices, AI systems refine feeding schedules, track water quality parameters, and find anomalies, thereby minimizing resource wastage and environmental impact. The health of aquatic organisms is primary to sustainable aquaculture. AI-driven systems assess the biotic and ecological information to find out symptoms of illness at an initial phase. Aquaculture farmers can apply periodical corrections such as regulating water quality parameters or introducing treatments through detecting illness factors and risk issues, which can pause the spread of disease and reduce the commercial losses (Gladju et al., 2022). AI-powered models evaluate the ecological effects of aquaculture activities and detect tactics for justifying antagonistic properties. Assessing the past information and triggering different circumstances, these models enhance the site selection, reduce the territory devastation, and condense contamination with better waste management practices. AI and ML algorithms can improve the output efficacy in aquaculture system by constant learning and optimization (Mohale et al., 2024). Evaluating information on feed composition, development ranges, and ecological issues, these technologies adjust nursing methods, breeding programs, and stocking densities, leading to increased production while reducing the source contributions. AI-based forecasting evaluations assess the market drifts, customers likings and peripheral influences, such as meteorological conditions, to give aquaculture organizers a significant perception for policy-making. Predicting the need and rate variations, farmers can alternate the production lists, optimize production levels, and adjust their marketing plans to maximize profitability and sustainability. The integration of AI and ML into aquaculture practices holds a great impact in progressing sustainability, yield, and resilience in the phase of future tasks. Connecting the power of data-driven insights and mechanization, the aquaculture industry adjusts the source consumption, alleviates ecological effects, and confirms the long-term feasibility of this vital food production sector (Mandal & Ghosh, 2023). As the technical novelties and study progressively increases, AI and MI are poised to act a key role in structuring the forthcoming sustainable aquaculture.

Conclusion

To sum up, the aquaculture remains to grow with the progressions in technological skills, mostly by the integration of AI and ML. This combination has knowingly improved different features of aquaculture, from developing culture systems for best yield to promote active precaution measures against diseases. Furthermore, AI and ML have transformed the realization of nourishment and development designs in aquatic animals, leading to effective nursing practices that gain more production. The incorporation of AI and MI into aquaculture methods leads to the sustainable aquaculture, confirms the food safety, and reduces ecological effects. Concerning the health of our oceans and ecosystems, acceptance of these hi-tech progressions will be essential for facing the increasing needs for the aquatic goods.

References

Ahmed, N., & Turchini, G. M. (2021). Recirculating aquaculture systems (RAS): environmental solution and climate change adaptation. *Journal of Cleaner Production, 297,* Article 126604.

Allsopp, M., Santillo, D., & Dorey, C. (2013). Sustainability in aquaculture: present problems and sustainable solutions. *Ocean Yearbook Online, 27*(1), 291–322.

Arechavala-Lopez, P., Cabrera-Álvarez, M. J., Maia, C. M., & Saraiva, J. L. (2022). Environmental enrichment in fish aquaculture: a review of fundamental and practical aspects. *Reviews in Aquaculture, 14*(2), 704–728.

Authority, V. F. (2014). *Best practice environmental management guidelines for recirculating aquaculture systems.*

Azra, M. N., Okomoda, V. T., & Ikhwanuddin, M. (2022). Breeding technology as a tool for sustainable aquaculture production and ecosystem services. *Frontiers in Marine Science, 9,* Article 679529.

Dawood, M. A. (2021). Nutritional immunity of fish intestines: important insights for sustainable aquaculture. *Reviews in Aquaculture, 13*(1), 642–663.

Drozdz, D., Malinska, K., Mazurkiewicz, J., Kacprzak, M., Mrowiec, M., Szczypiór, A., Postawa, P., & Stachowiak, T. (2020). Fish pond sediment from aquaculture production—current practices and the potential for nutrient recovery: a review. *International Agrophysics, 34*(1).

Engle, C. R., Kumar, G., & van Senten, J. (2020). Cost drivers and profitability of US pond, raceway, and RAS aquaculture. *Journal of the World Aquaculture Society, 51*(4), 847–873.

FAO. (2020). *The State of Food and Agriculture 2020. Overcoming water challenges in agriculture*. Food and Agriculture Organization of the United Nations.

Gladju, J., Kamalam, B. S., & Kanagaraj, A. (2022). Applications of data mining and machine learning framework in aquaculture and fisheries: a review. *Smart Agricultural Technology, 2*, Article 100061.

Kandathil, R. D., AkbarAli, I., Schmidt, B. V., John, E. M., Sivanpillai, S., & Thazhakot Vasunambesan, S. (2020). Improvement of nutritional quality of live feed for aquaculture: an overview. *Aquaculture Research, 51*(1), 1–7.

Khanjani, M. H., Zahedi, S., & Mohammadi, A. (2022). Integrated multitrophic aquaculture (IMTA) as an environmentally friendly system for sustainable aquaculture: functionality, species, and application of biofloc technology (BFT). *Environmental Science and Pollution Research, 29*(45), 67513–67531.

Lall, S. P., & Kaushik, S. J. (2021). Nutrition and metabolism of minerals in fish. *Animals, 11*(09), 2711.

Lieke, T., Meinelt, T., Hoseinifar, S. H., Pan, B., Straus, D. L., & Steinberg, C. E. (2020). Sustainable aquaculture requires environmental—friendly treatment strategies for fish diseases. *Reviews in Aquaculture, 12*(2), 943–965.

Mansour, A. T., Ashour, M., Alprol, A. E., & Alsaqufi, A. S. (2022). Aquatic plants and aquatic animals in the context of sustainability: cultivation techniques, integration, and blue revolution. *Sustainability, 14*(6), 3257.

Mandal, A., & Ghosh, A. R. (2023). Role of artificial intelligence (AI) in fish growth and health status monitoring: a review on sustainable aquaculture. *Aquaculture International, 10*, 1–30.

Mohale, H. P., Narsale, S. A., Kadam, R. V., Prakash, P., Sheikh, S., Mansukhbhai, C. R., Kirtikumar, P. B., & Baraiya, R. (2024). Artificial intelligence in fisheries and aquaculture: enhancing sustainability and productivity. *Archives of Current Research International, 24*(3), 106–123.

Morro, B., Davidson, K., Adams, T. P., Falconer, L., Holloway, M., Dale, A., Aleynik, D., Thies, P. R., Khalid, F., Hardwick, J., & Smith, H. (2022). Offshore aquaculture of finfish: big expectations at sea. *Reviews in Aquaculture, 14*(2), 791–815.

Mugimba, K. K., Byarugaba, D. K., Mutoloki, S., Evensen, Ø., & Munang'andu, H. M. (2021). Challenges and solutions to viral diseases of finfish in marine aquaculture. *Pathogens, 10*(6), 673.

Nasr-Eldahan, S., Nabil-Adam, A., Shreadah, M. A., Maher, A. M., & El-Sayed Ali, T. (2021). A review article on nanotechnology in aquaculture sustainability as a novel tool in fish disease control. *Aquaculture International, 29*, 1459–1480.

Nederlof, M. A., Verdegem, M. C., Smaal, A. C., & Jansen, H. M. (2022). Nutrient retention efficiencies in integrated multi—trophic aquaculture. *Reviews in Aquaculture, 14*(3), 1194–1212.

Ogello, E. O., Outa, N. O., Obiero, K. O., Kyule, D. N., & Munguti, J. M. (2021). The prospects of biofloc technology (BFT) for sustainable aquaculture development. *Scientific African, 14*, Article e01053.

Rajalakshmi, M., Manoj, V. R., & Manoj, H. (2022). Comprehensive review of aquaponic, hydroponic, and recirculating aquaculture systems. *Journal of Experimental Biology and Agricultural Science, 10*(6), 1266–1289.

Salin, K. R., Subasinghe, R. P., Senarathna, D., & Shinn, A. P. (2023). Cage culture of finfish: its importance, distributions and future modifications in

ongoing climate change. *Climate Change on Diseases and Disorders of Finfish in Cage Culture, 27*, 1–33.
Selvaraj, S., Chidambaram, P., Ezhilarasi, V., Kumar, P. P., Samuel Moses, T. L., Antony, C., & Ahilan, B. (2021). A review on the reproductive dysfunction in farmed finfish. *Annual Research & Review in Biology, 36*(10), 65–81.
Subasinghe, R., Alday-Sanz, V., Bondad-Reantaso, M. G., Jie, H., Shinn, A. P., & Sorgeloos, P. (2023). Biosecurity: reducing the burden of disease. *Journal of the World Aquaculture Society, 54*(2), 397–426.

2

Technological update in aquaculture

Introduction

Aquaculture—a fancy word for farming fish, shellfish, and seaweed, all that underwater jazz—has honestly become a big deal everywhere. It is not just about cranking out more shrimp cocktails for the masses; people actually care about how it is done these days. Ther is this whole vibe shift toward doing it smarter, greener, and, well, not trashing the planet in the process. That is where Integrated multitrophic aquaculture, or IMTA, if you are into acronyms, comes in. It is basically like setting up a roommate situation for different species—think fish, shellfish, and seaweed—where everyone's leftovers become someone else's lunch (Fig. 2.1). This means less waste and more use of all those nutrients that would otherwise just float around causing trouble (Chopin, 2006). Recirculating aquaculture systems (RASs) are like the Teslas of fish farming—closed-loop, super-efficient, barely any water wasted, and you do not end up with a swamp of fish poop at the end (Gupta et al., 2024). Honestly, it is wild how far things have come from the old-school fishponds. The industry's not perfect, but at least there is some actual effort to keep things sustainable, which is more than you can say for a lot of food production.

Automation and fancy sensors use technology to keep tabs on everything from water temperatures to fish feeding habits, and even to spot if one of them looks sick. Basically, it is like Fitbits for fish. All of this is supposed to help crank out more seafood with fewer screw-ups. Now, if you peek into the crystal ball, the future of aquaculture is both wild and a tad nerve-wracking. We are looking at almost 10 billion people by 2050. That is a heck of a lot of mouths to feed, and fish is only going to get more popular. So, the pressure is really on to level up and dream up new ways to keep the seafood coming without totally trashing the ocean. There is a bunch of work going into making fish food from bugs and plants instead of just grinding

Aquaculture with Artificial Intelligence. https://doi.org/10.1016/B978-0-443-33704-8.00005-7

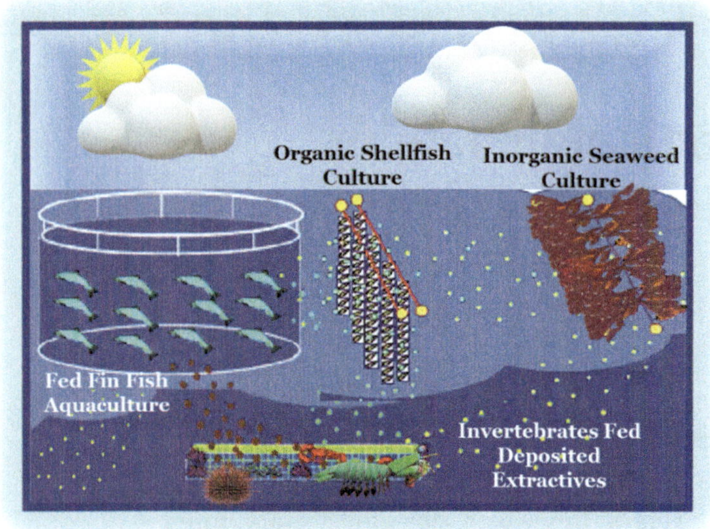

Figure 2.1 Integrated multitrophic aquaculture (IMTA) combining different species, such as fish, shellfish, and seaweeds.

up more wild fish. The idea is to stop raiding the oceans for feed and get a bit smarter about it. Societies are talking about taking fish farms offshore—like, way out into deep water. Imagine fish condos floating in the middle of nowhere. Mother nature is not always in the mood to cooperate. Bottom line: if we keep pushing tech, teaming up across industries, and not being total idiots about the environment, aquaculture could really save our butts in the food department. This chapter is all about those next-level aquaculture tricks, so farmers can keep up and, hopefully, not lose their shirts.

Technological update

Aquaculture has seen significant technological advancements; in fact, it is truly amazing how much technology has completely changed the fish farming industry. In the past, people would manually toss in the feed and cross their fingers, hoping the fish would not become overfed. These sophisticated automated feeders with sensors and intelligent software essentially look after your fish for you in this situation. They essentially know when your salmon are feeling snacky before you do by monitoring the water temperature, fish size, and even their movement (Fig. 2.2). Getting nuts is the genetic side of things.

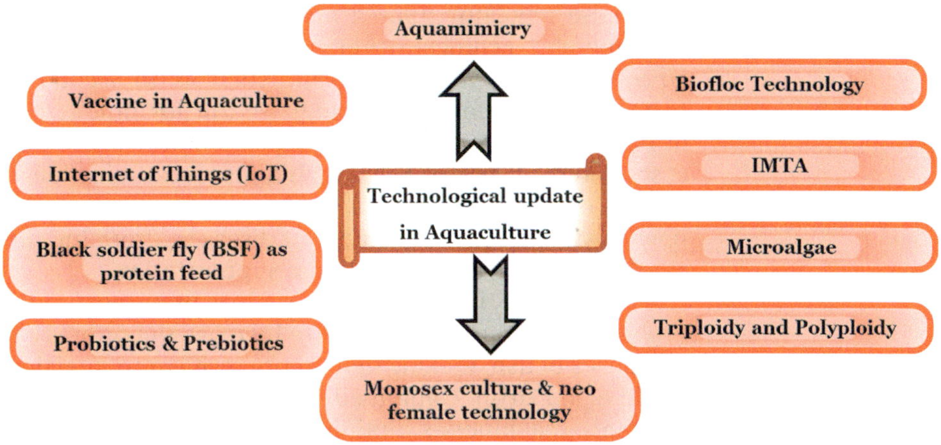

Figure 2.2 The most recent technologies updated in aquaculture.

We are discussing fish that require less effort to reach market size, grow more quickly, and are more resilient to illnesses. Fish that are healthier and require fewer antibiotics are the result of all that advanced DNA technology, including genome editing, genetic markers, and selective breeding. No more pouring a ton of chemicals into the water to keep everyone alive. This makes everything much more sustainable, which is important when you are trying to feed a planet that keeps producing more people, let's face it (Houston et al., 2020).

Additionally, developments in cryopreservation practices nowadays permit for the long-term storing of fish gametes, safeguarding biodiversity protection, and offering plasticity in propagation agendas. Quantified water is a perilous factor in aquaculture, as unhygienic water circumstances pave the way for disease eruptions, deprived fish development, and elevated death rates. Current technological progressions in water excellence supervising have permitted aquaculture agriculturalists to monitor up-to-date statistics on perilous variables, comprising oxygen levels, pH, temperature, and ammonia concentrations. Sensing detectors and remote-observing gears are currently combined into aquaculture organizations, allowing nonstop investigation of water circumstances. The above-mentioned schemes and systems alert agriculturalists of various deviations from the optimum assortment, permitting them to take rapid remedial schedules. Approximately of these schemes also have analytical competences, exploiting machine learning algorithms to estimate probable difficulties before they befall, which meaningfully decreases the hazard of disease eruptions. The problem of disease regulator ruins a noteworthy encounter in aquaculture; merely, technical novelties are making paces in this part. Unconventional analytical gears, such as molecular methods (e.g., PCR and gene sequencing), permit for timely recognition of pathogens, safeguarding appropriate interference before diseases blowout across fish inhabitants. Immunization machineries have also progressed, with additional active injections now existing for an assortment of diseases, dropping the necessity for antibiotics and additional suppositories.

Furthermore, novel methods, such as immunostimulants and probiotic flora, are being hired to increase the ordinary resistant responses of fish, making them more resilient to infections (Merrifield et al., 2010). Combined with better-quality biosecurity trials, these machineries diminish the occurrence of illnesses and the accompanying financial losses in aquaculture. Additionally, the expansion of substitute fish forages is a distinguished technical apprize expected to drop the ecological footprint of aquaculture. Conventional fish feeds depend deeply on fishmeal and fish oil derivative from wild-caught fish, which increases anxieties about overfishing. Novelties in feed expertise have led to the expansion of maintainable substitutes; for instance, feeds prepared from algae, insects, and plant-based proteins. The above said substitute feeds not only decrease burden on wild fish frameworks but also offer nutritious aids to cultivated fish (Fantatto et al., 2024). Scientific progressions in aquaculture are acting a vital role in shaping the industry's future. Since automatic feeding systems and genomic enhancements were playing a major part in advanced water quality supervising gears and disease controlling acquaintances, these developments are making aquaculture more efficient, sustainable, and moneymaking. Although technology endures to evolve, aquaculture is well-positioned to become a crucial player in universal food security, offering a supportable solution to the rising demand for protein.

Precision aquaculture

Meticulousness aquaculture is a ground-breaking tactic that affects the philosophies of accuracy cultivation to fish farming, merging innovative knowledge, statistics analytics, and mechanization to enhance the manufacture process. While universal petition for seafood upsurges due to inhabitants development and improved responsiveness of the vigor benefits of fish, accuracy aquaculture occurs as a crucial resolution to safeguarding maintainable and competent production. The impression centers on the usage of real-time data, devices, artificial intelligence (AI), and other innovative technologies to improve decision-making and progress the organization of aquaculture systems (Fore et al., 2018). In conventional aquaculture, fish agriculturalists regularly differ on practice, comment, and labor-rigorous practices to accomplish their processes. Nevertheless, these tactics are ineffective, labor-intensive, and prone to human error. Meticulousness aquaculture, in divergence, influences numerical technologies to supervise frequent environmental parameters and biotic situations with superior accuracy. Underwater sensing devices positioned in the water uninterruptedly supervise crucial variables, comprising water temperature, oxygen levels, pH, salinity, and nutrient concentrations. Hence, obtained instantaneous data are then administered by AI systems that distinguish templates, expect drifts, and recommend ideal comebacks, such as regulating feed rates or aeration systems. One of the greatest noteworthy aids of meticulousness aquaculture is the capability to display fish health and performance more meticulously. Traditional methods of disease detection in fish farms often comprise visual inspections, which are time-consuming and imprecise. In precision aquaculture, machineries, such as underwater cameras, biosensors, and image recognition software aid, distinguish initial signs of disease or anxiety in fish, authorizing farmers to take proactive actions before issues escalate. This diminishes the requirement for antibiotics and other chemicals, which not only drops production costs but also addresses anxieties about the environmental impact of aquaculture. Precision technologies have also renovated feeding practices in aquaculture. In conventional systems, overfeeding is a frequent problem that leads to wasted feed, increased production costs, and degradation of water quality due to uneaten food plummeting to the bottom (Quintanilla-Villanueva et al., 2023). Precision aquaculture uses automated feeders associated to sensors that detect fish behavior and hunger levels. These feeders deliver the meticulous quantity of food at the optimal time, diminishing waste and expanding growth rates. Through closely controlling feed distribution, farmers can recover feed conversion ratios and decrease environmental pollution, ultimately contributing to more sustainable practices. The role of data analytics in meticulousness aquaculture is vital and cannot be exaggerated. The amalgamation of big data permits the examination of giant amounts of information collected from various sources, including sensors, historical records, and external data, such as weather forecasts. This data-driven method permits more knowledgeable decision-making, serving farmers to expect and manage risks, enhance resource use, and improve the overall efficiency. Predictive analytics, for example, forecast changes in water conditions, allowing farmers to regulate their management approaches in advance to alleviate credible negative effects on fish health or growth (Oliveira et al., 2024). Automation plays a vigorous role in precision aquaculture, helping to moderate employment prices and increase operational efficiency. Computerized systems clasp routine tasks, such as feeding, cleaning, and monitoring, freeing up farmers to focus on supplementary strategic features of the business. Robotic systems are also being commercialized to handle more multifaceted jobs, such as underwater inspections of fish cages or the application of treatments to sick fish. These developments document superior scalability, making it simpler for farmers to accomplish large-scale procedures with measlier resources. Additional promising feature of precision aquaculture is the use of onward-thinking breeding practices. Selective breeding programs that depend on genetic data help develop strains of fish that are further resistant to diseases, grow faster, and flourish in specific environmental conditions. This progresses efficiency and diminishes the need for chemical interventions. Precision aquaculture also streamlines the application of RASs, which reuse water within the farming system, thereby reducing the need for large volumes of freshwater and diminishing the risk of contamination from seepage expulsion (Badiola et al., 2012). Nevertheless, while precision aquaculture offers plentiful benefits, it also grants challenges to its extensive implementation. One of the main barriers is the high cost of applying the essential technologies, predominantly for small- and medium-sized farms. The installation of sensors, automation systems, and AI-driven software needs notable upfront asset. Moreover, the impairment of dealing and transformation large datasets necessitate farmers to attain innovative technical skills or hire specialized personnel to assist them. In the path of overwhelming these challenges, teamwork between governments, research institutions, and the private sector is vital to afford judicious solutions, training programs, and financial support to fish farmers. Additional challenge is the necessity for consistent internet connectivity in remote coastal or rural areas where many fish farms are positioned. Meticulousness aquacultures depend deeply on cloud computing and data broadcast, and unpredictable internet access delays real-time data collection and investigation. Efforts to progress infrastructure in these areas are essential to safeguard that the benefits of precision aquaculture are manageable to all farmers, regardless of their location. As the aquaculture industry continues to progress, precision aquaculture will likely play a central role in guaranteeing its sustainability and efficiency.

Recirculating Aquaculture Systems

RASs represent a highly advanced and justifiable approach to fish farming, tackling many of the ecological and operative contests faced by traditional aquaculture practices (Badiola et al., 2018). Contrasting conservative systems, which trust deeply on a continuous supply of water from natural bases, such as rivers, lakes, or oceans, RAS is a closed-loop system premeditated to diminish water usage and ecological impact. It formulates this by successively cleaning and reprocessing the same water, henceforth the term "recirculating." This not only preserves water but also delivers better control over the aquatic environment, making it an increasingly popular choice for aquaculture production worldwide (Ende et al., 2024). In an illustrative RAS setup, fish are raised in tanks, and the water from these tanks is uninterruptedly treated to eradicate waste products, such as ammonia, nitrites, and carbon dioxide. The treatment process encompasses numerous steps, comprising mechanical filtration to remove solid particles, biological filtration to convert harmful ammonia into less toxic nitrates, and oxygenation to tolerate acceptable levels of dissolved oxygen for the fish. Cautiously supervising these parameters, RAS permits optimal circumstances for fish growth, succeeding in advanced survival rates, faster growth, and improved fish health compared with traditional systems.

Furthermore, the bolted nature of the organization declines the risk of disease incidences, as fish are not exposed to external pathogens or contaminants that may be existing in natural water bodies. One among the main key profits of RAS is its capability to stimulate in locations where traditional aquaculture is not possible, such as domestic areas far from natural water sources or provinces with limited water availability. This paves the way for new potentials for fish farming in urban environments or parched areas, making aquaculture more geologically assorted and obtainable (Ahmed & Turchini, 2021). Besides, the severe water supplies and waste expulsion make RAS a more ecologically approachable assortment, helping to alleviate the ecological influences regularly associated with traditional aquaculture, such as nutrient contamination, habitat degradation, and the spread of aggressive species. Substitute assistance of RAS is the high degree of regulator it bids over water quality and temperature, which allow farmers to enhance circumstances for specific species. This is predominantly vital for elevated-value species that require explicit ecological conditions for perfect progress, on behalf of instance salmon, trout, or shrimp. In an RAS, parameters, such as water temperature, pH, and salinity, were cautiously measured, safeguarding that the fish were continuously in a superlative environment for growth.

Moreover, the system's capability to reutilize water diminishes the requirement for huge extents of water, permitting farmers to harvest fish in areas with inadequate water resources, and it also decreases the hazard of disease transmission from the exterior environment. Yet, in spite of its numerous benefits, RAS also originates with precise encounters, predominantly in terms of its extraordinary original setup charges and energy utilization. The innovative purification and water treatment apparatus essential for RAS remained luxurious to setup and maintain, making it less available for moderate farmers or individuals in emerging areas with inadequate economic capitals. Furthermore, the system's dependence on electrical energy to power pumps, filters, and other apparatus means that energy costs were significant, particularly in areas where electricity prices are elevated. On the other hand, as expertise advances and renewable energy sources become further extensively obtainable, the functioning charges of RAS are predictable to decline, making the system more economically feasible in the long term (Bergman et al., 2020). Suggestively dropping water habit, improving fish health, and diminishing ecological influence, RAS offers a maintainable and well-organized technique to encounter the rising universal demand for seafood.

Genetic improvement

Genetic enhancement, frequently mentioned to as inherited improvement or genetic engineering, is the progression of adjusting an organism's genetic makeup to improve necessary personalities. This perception has played a vital part in farming, animal husbandry, and medication for generations. Genetic development through selective breeding and biotechnology is one of the furthermost active approaches for accomplishing these goals. Across the previous few eras, progressions in genetics have contributed meaningfully to refining development rates, disease resistance, feed efficiency, and environmental flexibility in numerous aquaculture species. The basis of inherited development in aquaculture lies in selective breeding (Fig. 2.3). This traditional technique encompasses selecting individuals

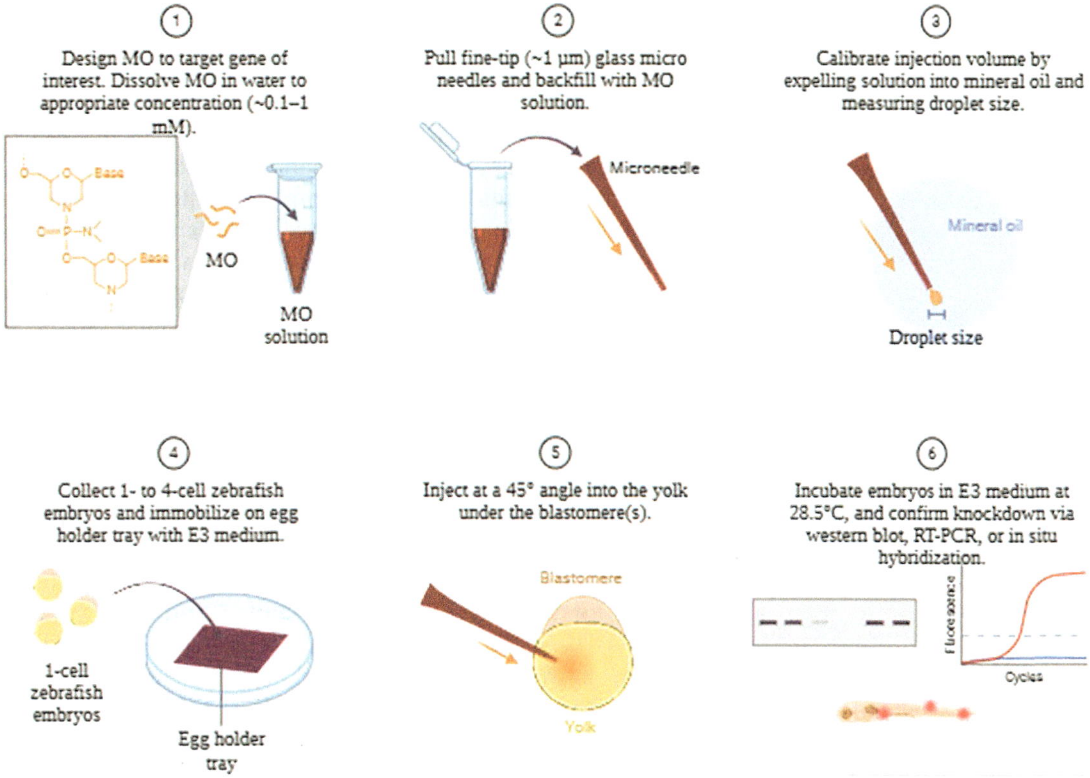

Figure 2.3 Mechanism of gene knockdown in zebra fish.

with essential qualities, for instance, faster growth, defiance to disease, or higher reproductive success, and breeding them over successive generations. Selective breeding has been extremely efficacious in species, such as Atlantic salmon, tilapia, and shrimp, resultant in noteworthy productivity developments. For example, the growth rates of selectively bred Atlantic salmon have improved by up to 10% per generation, subsequent in quicker period to market and additional well-organized resource habit. One among the alternative traditional tactic hybridization is where two different species or strains are crossed to produce offspring with better-quality personalities. Hybrids display heterosis, also recognized as hybrid vigor, where the offspring outperforms together parent species in terms of development, existence, or anxiety tolerance. Hybrid tilapia, for instance, have been extensively accepted in numerous countries for their superior growth rates and ecological adaptableness associated with their parental class (Yanez et al., 2015).

Beside improvements in molecular biology, inherent enhancement has progressed beyond traditional breeding practices. Marker-assisted selection (MAS) is a modern breeding method that exploits molecular markers and precise DNA sequences accompanying with necessary qualities to recognize and select individuals with the greatest hereditary possible. This technique permits further specific and quicker breeding associated to traditional assortment, as it does not rely exclusively on noticeable characters that ecological factors can influence. In aquaculture, MAS has been used to recover disease resistance, growth, and feed efficiency in species, such as Atlantic salmon, rainbow trout, and shrimp. Through integrating molecular markers into breeding agendas, aquaculture creators might attain hereditary improvements further rapidly and competently (Gjedrem et al., 2012). A more current progress in aquaculture genetics is genomic assortment, which includes information from the whole genome to envisage the breeding worth of individuals. Genomic assortment permits further precise predictions of multifaceted traits prejudiced by numerous inheritable factors, for instance, disease resistance and growth. This procedure has been predominantly effective in species such as Atlantic salmon, where it has been used to battle viral and bacterial diseases. Integrating genomic indication into breeding programs permits manufacturers to quicken hereditary enhancement and moderate the occurrence of disease eruptions, thereby improving the sustainability of aquaculture operations. Biotechnology has correspondingly played a vital part in hereditary enhancement, predominantly over the progress of transgenic species. Transgenic fish are hereditarily altered to fast exact traits that are not existing in their usual complements. For instance, transgenic salmon have been engineered to cultivate quicker by integrating a growth hormone inheritable factor from alternative fish species. Although transgenic fish have the probable to revolutionize aquaculture, their use ruins antagonistic remaining to concerns about ecological impacts and food fortification. Monitoring frameworks are still being industrialized in numerous countries to address these issues. Another significant feature of hereditary enhancement in aquaculture is the preservation and management of genetic properties. Cryopreservation, the progression of freezing and storing sperm, eggs, or embryos at very low temperatures, has become a treasured tool for conserving genetic diversity in aquaculture species. Germplasm banks, which stock these cryopreserved constituents, afford a protection in contradiction of the loss of genetic assortment due to discriminating breeding, disease eruptions, or environmental variations.

By preserving a repository of inherited material from diverse populations, germplasm banks help guarantee the long-term sustainability of aquaculture by permitting the restoration of genetic diversity into breeding agendas when desirable. Through selective breeding, molecular genetics, and biotechnology, aquaculture manufacturers have made noteworthy development in enlightening traits, such as growth, disease resistance, and environmental flexibility. As the field remains to progress, novel practices such as genomic assortment and the liable use of biotechnology will further accelerate genetic expansions (Song et al., 2023).

Alternative feeds

The examination of alternative feeds signifies a critical shift toward reducing dependency on wild-caught marine resources while preserving the nutritional integrity essential for optimal fish growth and health. One of the most auspicious alternative feed sources is the use of plant-based proteins, such as soy, corn, wheat, and peas. These constituents are extensively available, cost-effective, and have been shown to deliver adequate nourishment when formulated correctly. Soybean meal, in specific, has become one of the furthermost extensively used plant-based protein sources in aquaculture. It comprises a comparatively high protein content and is administered to eliminate antinutritional aspects, making it appropriate for an extensive variety of aquatic species. Nevertheless, the use of plant-based proteins presents several challenges. Fish, particularly carnivorous species, have progressed to digest animal-based proteins more competently, so regulating their diets to include plant-based substitutes requires careful formulation to safeguard that amino acid profiles, digestibility, and palatability are sustained. In addition, plant-based feeds occasionally comprise antinutritional compounds that could affect fish health and growth, necessitating further refinement in processing and feed development. Furthermore, to plant-based proteins, insect-based feeds have expanded consideration as a probable substitute. Insects are obviously consumed by many fish species in the wild, and they are shaped sustainably using organic waste materials, dropping the environmental footprint of feed invention. Black soldier fly larvae (BSFL) are one of the maximum auspicious insect species for aquaculture feed due to their high protein and fat content. BSFL were cultivated professionally and on a large scale, making them a feasible assortment for reducing reliance on fishmeal (Maulu et al., 2022). Studies have confirmed that fish-fed insect-based diets accomplish comparable growth rates and health consequences as those fed traditional fishmeal-based diets. Nevertheless, trials, such as scaling production, consumer receipt, and monitoring consent, need to be addressed before insect-based feeds are widely adopted in commercial aquaculture. Alternative feed ingredient gaining adhesion is algae. Algae, comprising both microalgae and macroalgae (seaweed), are nutrient-rich and can be cultured in marine environments without the need for arable land or freshwater, making them a biologically favorable option. Algae are high in vital fatty acids, proteins, and vitamins, and certain species of algae even be rich in omega-3 fatty acids, which are critical for fish health and are typically obtained from fish oil (Zhang et al., 2024). The use of algae in aquaculture feeds not only reduces reliance on fishmeal and fish oil but also provides a sustainable foundation of important nutrients for fish growth and expansion. However, the making cost of algae, predominantly microalgae, remains a noteworthy barricade to its extensive practice. Improvements in cultivation technology and administering approaches are needed to diminish costs and make algae-based feeds economically inexpensive. Fermentation-based feeds are alternative promising avenue for the development of alternative feeds. Single-cell organisms, for instance, bacteria, yeast, and fungi, are grown on a diversity of substrates, including agricultural byproducts, and used to produce protein-rich biomass appropriate for aquaculture feeds. These microbial proteins offer numerous advantages: they are produced at scale, necessitate less land and water than traditional crops, and stipulate a consistent supply of nutrients. In specific, certain anxieties of bacteria and yeast are enriched with essential nutrients such as amino acids, vitamins, and omega-3 fatty acids, making them a highly nutritious option for aquaculture. Furthermore, fermentation procedures are tailored to specific nutrient requirements, offering flexibility in feed formulation. However, the cost of manufacture remains a challenge, as does optimizing fermentation processes to exploit nutrient yield and minimize waste. Alternative emerging feed source includes the utilization of agricultural and industrial byproducts, such as rice bran, distiller grains, and fish processing waste. These byproducts are often underused and repurposed as feed ingredients, reducing waste and lowering the cost of feed production. Fishmeal substitutes derived from fish processing waste, for example, have the potential to deliver a high-quality protein source while justifying the environmental impact of fish processing industries. The trial with using byproducts is safeguarding that they are of consistent quality and free from pollutants that could harm fish health. Proper processing and quality control are critical in developing these alternative feeds into safe and nutritious possibilities for aquaculture. The expansion and execution of alternative feeds in aquaculture have noteworthy implications for both the industry and the environment. Ecologically, dropping the reliance on wild-caught fish for feed production could help alleviate pressure on marine ecosystems and subsidize to the conservation of wild fish stocks. Economically, alternative feeds have the potential to alleviate feed prices by reducing dependence on volatile fishmeal and fish oil markets. Additionally, alternative feeds, predominantly those obtained from insects, algae, and fermentation processes, are produced locally, reducing transportation costs and promoting regional self-sufficiency in aquafeed production. Though, the implementation of alternative feeds is not deprived of its challenges. One of the chief alarms is guaranteeing that alternative feeds provide the same level of nutrition as traditional feeds. Aquatic species have specific dietary requirements, particularly for essential amino acids, fatty acids, and micronutrients, which must be met to preserve optimal growth, well-being, and reproductive performance. Slightly deficiencies in these nutrients lead to reduced growth rates, compromised immune function, and lower overall production efficiency. Consequently, alternative feed formulations must be sensibly industrialized to confirm that they meet the nutritional needs of specific species. Continued research and invention in feed formulation, processing technologies, and production methods are vital to overcome existing limitations and facilitating the extensive adoption of alternative feeds in the aquaculture industry. Embracing these alternatives, the aquaculture segment moves toward a more maintainable and hardy future, ensuring the long-term feasibility of global fish production while minimizing its ecological footprint.

Automation and robotics

As the world population grows and wild fish stocks diminish due to overfishing and environmental changes, aquaculture bids a justifiable solution for meeting the increasing demand for seafood. Nevertheless, traditional aquaculture practices face numerous challenges, comprising labor-rigorous operations, high operational costs, and environmental influences. Automation and robotics are transforming aquaculture by addressing these challenges and permitting more effective, sustainable, and scalable production methods. The key assistances of automation in aquaculture are the capability to enhance feeding practices. In traditional aquaculture, feeding is often manual, which leads to overfeeding or underfeeding, damagingly affecting fish health and water quality. Automated feeding systems, equipped with sensors and artificial intelligence (AI), are designed to deliver the right amount of feed at the right time. These systems monitor environmental conditions, such as water temperature, oxygen levels, and fish behavior, to adjust feeding agendas and portions, thereby minimizing waste and safeguarding optimal growth. This level of precision not only diminishes feed costs, which account for up to 50% of operational expenses, but also improves the sustainability of the operation by minimizing nutrient discharge into surrounding water bodies (Luna et al., 2016). Robotics is also playing a vital role in refining the observing and preservation of aquaculture environments. Underwater drones and autonomous vehicles equipped with cameras and sensors are now used to monitor fish health, water quality, and infrastructure, such as nets and cages. These robotic systems activate in interesting conditions, such as deep water or areas with strong currents, where human divers might find it difficult or dangerous to work. Real-time data collected by these robots allow aquaculture farmers to detect issues, such as disease outbreaks, equipment damage, or water quality deterioration before they become critical. This proactive approach to monitoring helps improve the overall health and yield of aquaculture operations while reducing the need for human intervention (Bohara et al., 2024). Furthermore, robotics has made noteworthy strides in automating fish gathering processes. Conventionally, collecting fish is labor-intensive, demanding manual handling that might be stressful for both the employees and the fish. Robotic harvesters, however, efficiently capture, sort, and transport fish, suggestively reducing labor costs and improving animal welfare by minimizing handling anxiety. These machines are combined with AI configuration that analyze fish size and weight to safeguard that only mature fish are harvested, permitting younger stock to continue growing and augmenting long-term sustainability and yield. Furthermore, to these progressions, the incorporation of mechanization and robotics in aquaculture supports the development of smart farms. These farms are equipped with interconnected systems that enable real-time data collection, examination, and decision-making. Smart aquaculture farms influence the Internet of Things (IoT), cloud computing, and machine learning algorithms to generate a self-regulating environment. For occasion, water circulation systems are automated to sustain optimal oxygen levels, while temperature control systems adjust according to seasonal fluctuations (Von Borstel et al., 2013). The continuous flow of data from sensors disseminated throughout the farm affords appreciated insights into patterns and trends, permitting prognostic maintenance and operational adjustments that enhance efficiency and minimize downtime. Sustainability is a serious anxiety in aquaculture, and automation offers solutions that alleviate ecological influences. Reducing human intervention, automated systems limit the risk of contamination and disease transmission between fish and workers. Robotics also assist in cleaning and maintaining aquaculture infrastructure, such as tanks and nets, preventing the accumulation of harmful biofouling organisms. Furthermore, AI-driven systems track and decrease the use of antibiotics and other chemicals, indorsing healthier ecosystems and safer seafood products. Automatic waste management systems that gather and treat organic waste from fish farms further contribute to minimizing environmental pollution. The economic implications of automation and robotics in aquaculture are profound. Reducing labor budgets, improving efficiency, and minimizing waste, these technologies offer noteworthy cost savings for aquaculture businesses. Additionally, automation permits scalability, allowing small- and medium-sized enterprises to enlarge their processes deprived of equivalently cumulative labor and resource contributions. This scalability is serious in regions where aquaculture is a key driver of economic growth and food security (Wu et al., 2022). Automation also unlocks up new opportunities for high-tech job creation as the demand for skilled workers to accomplish and maintain these advanced systems grows. However, the widespread adoption of automation and robotics in aquaculture is not deprived of challenges. The elevated initial investment costs related with purchasing and integrating these technologies are a barrier for smaller operations. There is also a need for specialized training and education for aquaculture workers to effectively operate and maintain automated systems. Moreover, as with any technological progression, concerns arise about potential job displacement in regions where aquaculture employs large numbers of people. Addressing these challenges will necessitate collaboration between industry stakeholders, governments, and educational institutions to progress policies and programs that support the transition to automated aquaculture. As the demand for seafood endures to grow, the adoption of automation and robotics will play an essential role in ensuring that aquaculture meets this demand while minimizing environmental influences and promoting economic growth. The future of aquaculture lies in embracing these innovations to create a more resilient and sustainable blue economy (Lim, 2024).

Integrated multitrophic aquaculture

IMTA involves the coculture of numerous species in the identical aquatic environment, where one species exploits the byproducts or waste of another species. This integrated approach enhances resource utilization, minimizes environmental impact, and progresses overall system resilience (Loayza-Aguilar et al., 2023).

Blockchain and traceability

Blockchain technology is being explored to recover traceability and transparency in the aquaculture source chain (Ismail et al., 2023). Recording production and spreading data on a tamper-proof ledger, blockchain aids authenticate the cause and eminence of seafood products, thereby augmenting consumer faith and food security. These technological apprizes imitate ongoing efforts to make aquaculture more maintainable, well-organized, and resilient in the face of developing environmental and market challenges (Luna et al., 2024). However, it is indispensable to note that the landscape of aquaculture technology endures to grow, with ongoing research and development driving further innovations in the field.

Conclusion

Sustainable aquaculture is a world-wide requirement at present. This chapter has emphasized the rising inventions that push the aquaculture industry toward sustainability and productivity. Tools such as nutrient-rich feed alternatives, automation, and robotics in aquaculture demonstrate significant milestones in technological progression. These tools progress productional effectiveness and decrease distressing impacts on both the environment and fish health. Ecosystem-based farming practices are enhanced by integrated multitrophic aquaculture (IMTA); however, transparency and consumer faith are strengthened by blockchain technology. Nevertheless, these innovations face challenges, such as high production costs, difficulty in technological adoption, and a lack of skilled personnel. To overcome these challenges, coordination among industry, academia, and policymakers should be developed. The future of the aquaculture industry will be determined by increasing the possibility of realizing the techniques mentioned above in sectors ranging from minor aquaculture farms to huge-scale production units.

References

Ahmed, N., & Turchini, G. M. (2021). Recirculating aquaculture systems (RAS): Environmental solution and climate change adaptation. *Journal of Cleaner Production, 297*, 126604.

Badiola, M., Basurko, O. C., Piedrahita, R., Hundley, P., & Mendiola, D. (2018). Energy use in recirculating aquaculture systems (RAS): A review. *Aquacultural Engineering, 81*, 57–70.

Badiola, M., Mendiola, D., & Bostock, J. (2012). Recirculating aquaculture systems (RAS) analysis: main issues on management and future challenges. *Aquacultural Engineering, 51*, 26–35.

Bergman, K., Henriksson, P. J., Hornborg, S., Troell, M., Borthwick, L., Jonell, M., Philis, G., & Ziegler, F. (2020). Recirculating aquaculture is possible without major energy tradeoff: life cycle assessment of warmwater fish farming in Sweden. *Environmental Science & Technology, 54*(24), 16062–16070.

Bohara, K., Joshi, P., Acharya, K. P., & Ramena, G. (2024). Emerging technologies revolutionising disease diagnosis and monitoring in aquatic animal health. *Reviews in Aquaculture, 16*(2), 836–854.

Chopin, T. (2006). Integrated multitrophic aquaculture. *Northern Aquaculture, 12*(4), 4.

Ende, S., Henjes, J., Spiller, M., Elshobary, M., Hanelt, D., & Abomohra, A. (2024). Recent advances in recirculating aquaculture systems and role of microalgae to close system loop. *Bioresource Technology, 14*, Article 131107.

Fantatto, R. R., Mota, J., Ligeiro, C., Vieira, I., Guilgur, L. G., Santos, M., & Murta, D. (2024). Exploring sustainable alternatives in aquaculture feeding: the role of insects. *Aquaculture Reports, 37*, Article 102228.

Føre, M., Frank, K., Norton, T., Svendsen, E., Alfredsen, J. A., Dempster, T., Eguiraun, H., Watson, W., Stahl, A., Sunde, L. M., & Schellewald, C. (2018). Precision fish farming: a new framework to improve production in aquaculture. *Biosystems Engineering, 173*, 176–193.

Gjedrem, T., Robinson, N., & Rye, M. (2012). The importance of selective breeding in aquaculture to meet future demands for animal protein: a review. *Aquaculture, 350*, 117–129.

Gupta, S., Makridis, P., Henry, I., Velle-George, M., Ribicic, D., Bhatnagar, A., Skalska-Tuomi, K., Daneshvar, E., Ciani, E., Persson, D., & Netzer, R. (2024). Recent developments in recirculating aquaculture systems: a review. *Aquaculture Research, 2024*(1), Article 6096671.

Houston, R. D., Bean, T. P., Macqueen, D. J., Gundappa, M. K., Jin, Y. H., Jenkins, T. L., Selly, S. L., Martin, S. A., Stevens, J. R., Santos, E. M., & Davie, A. (2020). Harnessing genomics to fast-track genetic improvement in aquaculture. *Nature Reviews Genetics, 21*(7), 389–409.

Ismail, S., Reza, H., Salameh, K., Kashani Zadeh, H., & Vasefi, F. (2023). Toward an intelligent blockchain IoT-enabled fish supply chain: a review and conceptual framework. *Sensors, 23*(11), 5136.

Lim, L. W. (2024). Implementation of artificial intelligence in aquaculture and fisheries: deep learning, machine vision, big data, internet of things, robots

and beyond. *Journal of Computational and Cognitive Engineering, 3*(2), 112–118.
Loayza-Aguilar, R. E., Huamancondor-Paz, Y. P., Saldaña-Rojas, G. B., & Olivos-Ramirez, G. E. (2023). Integrated multitrophic aquaculture (IMTA): Strategic model for sustainable mariculture in Samanco Bay, Peru. *Frontiers in Marine Science, 10,* Article 1151810.
Luna, F. D., de la Rosa Aguilar, E., Naranjo, J. S., & Jagüey, J. G. (2016). Robotic system for automation of water quality monitoring and feeding in aquaculture shadehouse. *IEEE Transactions on Systems, Man, and Cybernetics: Systems, 47*(7), 1575–1589.
Luna, M., Fernandez-Vazquez, S., Castelao, E. T., & Fernández, Á. A. (2024). A blockchain-based approach to the challenges of EU's environmental policy compliance in aquaculture: from traceability to fraud prevention. *Marine Policy, 159,* Article 105892.
Maulu, S., Langi, S., Hasimuna, O. J., Missinhoun, D., Munganga, B. P., Hampuwo, B. M., Gabriel, N. N., Elsabagh, M., Van Doan, H., Kari, Z. A., & Dawood, M. A. (2022). Recent advances in the utilization of insects as an ingredient in aquafeeds: a review. *Animal Nutrition, 11,* 334–349.
Merrifield, D. L., Dimitroglou, A., Foey, A., Davies, S. J., Baker, R. T., Bøgwald, J., Castex, M., & Ringø, E. (2010). The current status and future focus of probiotic and prebiotic applications for salmonids. *Aquaculture, 302*(1–2), 1–8.
Oliveira, J., Oliva-Teles, A., & Couto, A. (2024). Tracking biomarkers for the health and welfare of aquaculture fish. *Fishes, 9*(7), 289.
Quintanilla-Villanueva, G. E., Maldonado, J., Luna-Moreno, D., Rodríguez-Delgado, J. M., Villarreal-Chiu, J. F., & Rodríguez-Delgado, M. M. (2023). Progress in plasmonic sensors as monitoring tools for aquaculture quality control. *Biosensors, 13*(1), 90.
Song, H., Dong, T., Yan, X., Wang, W., Tian, Z., Sun, A., Dong, Y., Zhu, H., & Hu, H. (2023). Genomic selection and its research progress in aquaculture breeding. *Reviews in Aquaculture, 15*(1), 274–291.
Von Borstel, F. D., Suárez, J., de la Rosa, E., & Gutiérrez, J. (2013). Feeding and water monitoring robot in aquaculture greenhouse. *Industrial Robot: An International Journal, 40*(1), 10–19.
Wu, Y., Duan, Y., Wei, Y., An, D., & Liu, J. (2022). Application of intelligent and unmanned equipment in aquaculture: a review. *Computers and Electronics in Agriculture, 199,* Article 107201.
Yáñez, J. M., Newman, S., & Houston, R. D. (2015). Genomics in aquaculture to better understand species biology and accelerate genetic progress. *Frontiers in Genetics, 6,* 128.
Zhang, Z., Miar, Y., Huyben, D., & Colombo, S. M. (2024). Omega—3 long—chain polyunsaturated fatty acids in Atlantic salmon: functions, requirements, sources, de novo biosynthesis and selective breeding strategies. *Reviews in Aquaculture, 16*(3), 1030–1041.

3

Smart and digitalizing the aquaculture

Introduction

Aquaculture, the farming of aquatic organisms, including fish, crustaceans, mollusks, and aquatic plants, has become one of the fastest-growing food production sectors globally, contributing significantly to food security, nutrition, and economic development. With the world's population projected to exceed 9 billion by 2050, sustainable food systems are more critical than ever. Traditional aquaculture, although productive, is often hindered by issues such as disease outbreaks, environmental degradation, resource inefficiencies, and a lack of real-time monitoring. In response to these challenges, the integration of smart and digital technologies in aquaculture has emerged as a transformative approach to revolutionize the sector through enhanced productivity, sustainability, and traceability (Bostock et al., 2010; FAO, 2020). Smart aquaculture refers to the application of information and communication technologies (ICT), the Internet of Things (IoT), artificial intelligence (AI), big data analytics, blockchain, robotics, and remote sensing to monitor, analyze, and optimize aquaculture operations in real time. Digitalization enables farmers and stakeholders to make data-driven decisions, improve efficiency, predict and prevent risks, and ensure compliance with environmental and regulatory standards. This transition from traditional to smart aquaculture marks a significant shift toward Industry 4.0 in aquatic food production systems (Lezzi et al., 2018; Verdouw et al., 2016). The integration of IoT in aquaculture systems enables real-time monitoring of water quality parameters, including temperature, dissolved oxygen, salinity, pH, and ammonia levels, which are essential for the health and growth of aquatic organisms. Wireless sensor networks (WSNs) deployed in ponds, tanks, or cages enable continuous data collection, transmission, and cloud-based storage, providing actionable insights for farmers to optimize feeding regimes, manage stocking densities, and control environmental conditions (Chaudhary

Aquaculture with Artificial Intelligence. https://doi.org/10.1016/B978-0-443-33704-8.00011-2

et al., 2022). These digital tools reduce manual labor and enable early detection of potential issues, thus minimizing losses because of diseases or poor water quality. AI and machine learning (ML) algorithms are increasingly being applied to analyze complex datasets generated from aquaculture operations. AI-powered predictive models forecast growth rates, disease outbreaks, and feed conversion ratios, allowing proactive management. For example, image processing and computer vision are used for automatic fish counting, behavior monitoring, and health assessment, significantly improving accuracy and operational efficiency (Rojas-Torres et al., 2022). Moreover, smart feeding systems equipped with sensors and AI algorithms optimize feeding schedules and amounts based on real-time data, thereby reducing feed waste, one of the most significant operational costs in aquaculture, and improving water quality. Blockchain technology introduces transparency and traceability across the aquaculture value chain, from hatcheries to markets. Recording immutable data entries at each stage of production, processing, and distribution, blockchain ensures food safety, authenticity, and quality, which global consumers increasingly demand. This technology also strengthens supply chain integrity and facilitates compliance with international trade standards and certifications (Leng et al., 2018; Tian, 2016). Remote sensing and satellite technologies play a crucial role in the large-scale spatial monitoring of aquaculture farms, particularly in coastal and offshore environments. These technologies provide data on sea surface temperature, chlorophyll concentration, and the occurrence of harmful algal blooms, supporting site selection, environmental impact assessments, and risk mitigation strategies (Xiao et al., 2020). Drones are also being used for aerial surveillance, mapping, and biomass estimation in aquaculture facilities, enhancing the precision and coverage of farm management practices. The digital transformation of aquaculture holds immense potential for enhancing production efficiency, promoting environmental sustainability, and fostering socioeconomic resilience. However, the implementation of smart technologies also presents certain challenges, particularly in terms of cost, technical expertise, infrastructure, and data privacy. Capacity building, policy support, and collaboration among stakeholders, including governments, academia, industry, and farming communities, are crucial for fostering inclusive digital innovation in aquaculture (FAO, 2022). In conclusion, smart and digital aquaculture is not merely a technological advancement but a paradigm shift that redefines how aquatic food systems operate in the 21st century. Aquaculture has become more efficient, sustainable, and resilient, aligning with the goals of the Blue Economy and the United Nations Sustainable Development Goals (SDGs), particularly SDG 2 (zero hunger), SDG 9 (industry, innovation, and infrastructure), and SDG 14 (life below water). This book chapter explores the components, applications, benefits, and prospects of smart and digital aquaculture, offering a comprehensive framework for researchers, practitioners, and policymakers seeking to shape the future of aquatic food production.

IoT and sensor-based monitoring systems

The integration of the IoT and sensor-based monitoring systems has revolutionized the aquaculture industry, driving it toward precision farming, environmental sustainability, and enhanced productivity (Fig. 3.1). Traditional aquaculture practices, which relied heavily on manual labor and observational methods, are increasingly being replaced by intelligent systems capable of real-time monitoring, data collection, and remote management of aquaculture environments. IoT-enabled solutions are structured around interconnected sensors, devices, cloud computing platforms, and software analytics that collectively optimize operational efficiency and biosecurity in aquaculture systems (Zhao et al., 2019). At the core of smart aquaculture is the deployment of a network of sensors that continuously measure vital water quality parameters, including temperature, pH, dissolved oxygen (DO), ammonia, turbidity, and salinity. These

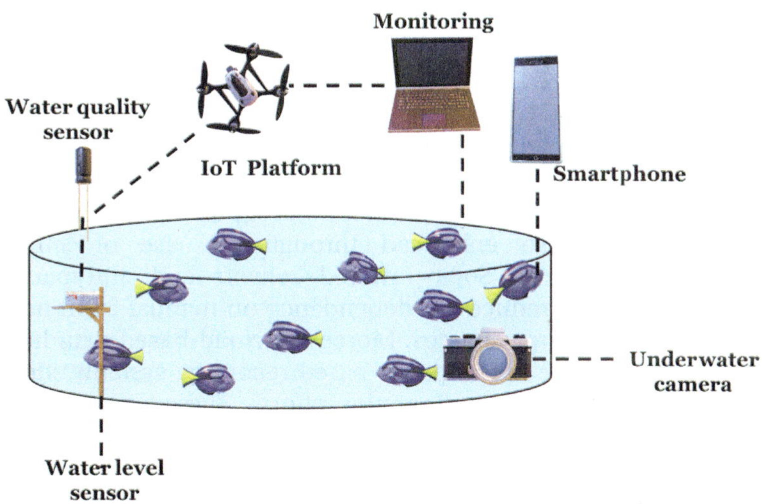

Figure 3.1 IoT and sensor-based monitoring systems in aquaculture.

parameters are critical as they directly influence the health, growth rate, feed efficiency, and overall survival of aquatic organisms. For instance, sensors embedded in ponds or tanks detect drops in dissolved oxygen levels, prompting automated aeration systems to activate and restore balance, thereby preventing fish stress or mortality (Bhattacharyya et al., 2021). Moreover, pH sensors provide early warning signals of water acidification, while ammonia sensors help monitor waste accumulation, which is particularly crucial in recirculating aquaculture systems (RAS).

IoT systems utilize wireless communication technologies, including Wi-Fi, LoRaWAN, Zigbee, and cellular networks, to transmit data from field sensors to centralized cloud platforms or mobile applications in real time. This remote accessibility allows farm operators, researchers, and stakeholders to make data-driven decisions regardless of their physical location. Advanced platforms integrate AI and ML algorithms that analyze the collected data to detect trends, forecast potential hazards, and suggest adaptive responses. Predictive analytics optimize feeding schedules, reducing waste and improving feed conversion ratios, which are major contributors to operational costs in aquaculture (Ramasamy & Vigneswaran, 2020). Another significant application of sensor-based IoT systems in aquaculture is biosurveillance. Wearable sensors and noninvasive imaging devices are attached to fish to monitor physiological metrics, stress indicators, or behavioral changes. Coupled with image processing systems and underwater cameras, these sensors provide real-time insights into fish activity, health anomalies, and aggressive interactions.

Such technologies not only help in disease outbreak detection but also facilitate prompt isolation or treatment, reducing the spread and impact of infections. Furthermore, the integration of remote sensing and GIS tools with IoT devices enables the monitoring of environmental parameters at broader spatial scales, particularly in cage culture, coastal, and offshore aquaculture environments (Zhou et al., 2021). Energy efficiency and sustainability are also enhanced through the use of smart aquaculture systems. Solar-powered sensors and automated feeding machines reduce the dependency on manual labor and nonrenewable energy sources. Moreover, cloud-based data logging minimizes the use of paper-based recording systems, supporting environmental conservation efforts. These systems also contribute to traceability and regulatory compliance by maintaining detailed digital logs of production and ecological parameters, aligning with the rising demand for transparency in seafood supply chains. Despite their numerous advantages, the implementation of IoT and sensor-based systems in aquaculture also poses challenges. These include the high initial capital investment, need for technical expertise, potential data security vulnerabilities, and sensor calibration or fouling in aquatic environments.

Nevertheless, with decreasing sensor costs, advancements in battery technology, and growing interest in digital aquaculture solutions, these barriers are gradually diminishing (Ahmed et al., 2020). In conclusion, IoT and sensor-based monitoring systems form the backbone of smart aquaculture, offering real-time visibility, proactive management, and sustainability. Their integration into aquaculture operations not only increases productivity but also ensures fish welfare, environmental health, and economic viability. As aquaculture continues to expand to meet the global demand for seafood, embracing these digital innovations is indispensable for creating resilient and future-ready aquaculture ecosystems.

Automation and robotics in aquaculture

The rapid advancement of automation and robotics has begun to significantly reshape the landscape of modern aquaculture, offering transformative solutions to challenges in productivity, sustainability, and labor efficiency. As the global demand for aquatic food products increases, particularly in light of dwindling wild fish stocks and environmental constraints, aquaculture must evolve toward more intelligent and integrated systems. Automation and robotics are essential components of this transition, enabling precision aquaculture practices that optimize resource utilization, reduce operational costs, and enhance animal welfare (Føre et al., 2018). One of the primary drivers for automation in aquaculture is the need to monitor and manage large-scale farming systems more effectively. Traditionally, aquaculture operations relied heavily on manual labor for tasks such as feeding, health monitoring, and water quality assessment. These processes are time-consuming, labor-intensive, and prone to human error. With the advent of automated systems, such as real-time sensors, feeding robots, and environmental monitoring platforms, farmers can now continuously assess critical parameters, including dissolved oxygen, pH, temperature, and turbidity. These smart sensors are often integrated with IoT platforms and cloud computing services, enabling remote access and control of aquaculture facilities (Zhao et al., 2021). Feeding, which accounts for a substantial portion of operational costs in aquaculture, has been significantly improved through the use of automation. Intelligent feeding systems employ data from underwater cameras, sensors, and AI algorithms to analyze fish behavior and determine optimal feeding times and quantities. This approach minimizes feed wastage, prevents overfeeding, and improves the feed conversion ratio, thus enhancing overall profitability. Automated feeders equipped with machine vision adaptively distribute feed across different zones in a cage or tank, addressing variability in fish distribution and ensuring more uniform growth (Hvas et al., 2020). Robotics has further extended the reach of automation into areas that are difficult or hazardous for human workers. Underwater robots, including remotely operated vehicles (ROVs) and autonomous underwater vehicles (AUVs), are increasingly employed for inspection, cleaning, and maintenance tasks. These robots perform tasks such as net cleaning, fish cage inspection, and biofouling removal, eliminating the need for divers and reducing safety risks and operational downtime. Some robotic systems are also equipped with high-resolution cameras and sonar to detect structural issues or monitor fish health indicators, such as lesion detection and abnormal swimming patterns (Skøien et al., 2019). Moreover, robotic arms and drones have found utility in hatcheries and onshore aquaculture systems. In hatcheries, automated egg handling, larval sorting, and vaccination machines streamline repetitive tasks while ensuring higher accuracy and sterility. Aerial drones, on the other hand, are used in pond aquaculture for aerial imaging and surveillance. These drones capture high-resolution, multispectral images that, when processed through AI models, provide insights into algal blooms, fish distribution, and variations in water quality. Such information is critical for proactive farm management and disease prevention (Mohanraj et al., 2020). Integrating automation with AI and ML has led to the development of predictive analytics tools that support decision-making in aquaculture. For instance, ML models trained on historical data predict optimal harvesting times, estimate biomass, and forecast disease outbreaks. When combined with robotics, these tools initiate automatic responses, such as triggering treatments or adjusting feeding schedules, without human intervention. This closed-loop control system is a cornerstone of the concept of "smart aquaculture," a paradigm in which interconnected devices and intelligent algorithms manage aquaculture operations with minimal human input (Li et al., 2021). Despite these technological advancements, challenges remain in the widespread adoption of automation and robotics in aquaculture. The initial cost of robotic systems, particularly for small and medium-scale farmers, remains high. Moreover, integrating various technologies into a cohesive system that is user-friendly and reliable under diverse environmental conditions requires interdisciplinary collaboration across engineering, biology, and data science. There is also a need for standardization and regulatory frameworks that ensure the safety and ethical deployment of autonomous systems in aquatic environments (Marr, 2020). Nevertheless, the future of automation and robotics in aquaculture is promising. As technologies mature and become more cost-effective, their adoption is expected to grow rapidly, particularly in countries with large-scale aquaculture industries, such as Norway, China, and India. Emerging technologies, such as swarm robotics, digital twins, and blockchain integration, are poised to revolutionize the sector further. Swarm robotics, inspired by the collective behavior of social animals, could enable coordinated tasks such as fish monitoring or cage maintenance using multiple small robots. Digital twins and virtual replicas of physical systems could enable real-time simulations and scenario testing, thereby enhancing precision and control. Meanwhile, blockchain could ensure traceability and transparency in automated supply chains (Tsolakis et al., 2021).

Artificial intelligence and data analytics

The integration of AI and data analytics (Fig. 3.2) has revolutionized the aquaculture industry, transitioning it from traditional practices to a more intelligent and sustainable model. AI enables machines to mimic human cognitive functions such as learning and decision-making, which, when combined with advanced data analytics, facilitates real-time monitoring, prediction, and optimization of aquaculture systems. In aquaculture, vast amounts of data are generated from sensors, IoT devices, satellite imaging, and farming equipment. AI algorithms, particularly ML and deep learning models, analyze these data to identify patterns, detect anomalies, and make informed decisions that enhance productivity and sustainability. For instance, AI-driven vision systems are utilized for automatic fish species identification, size estimation, and behavior monitoring, significantly reducing manual labor and improving accuracy (Zhao et al., 2021). Predictive analytics, based on historical and real-time data, forecast disease outbreaks, optimize feeding schedules, and manage water quality parameters such as pH, temperature, dissolved oxygen, and ammonia levels, ensuring healthier aquatic environments (Raghavan et al., 2020).

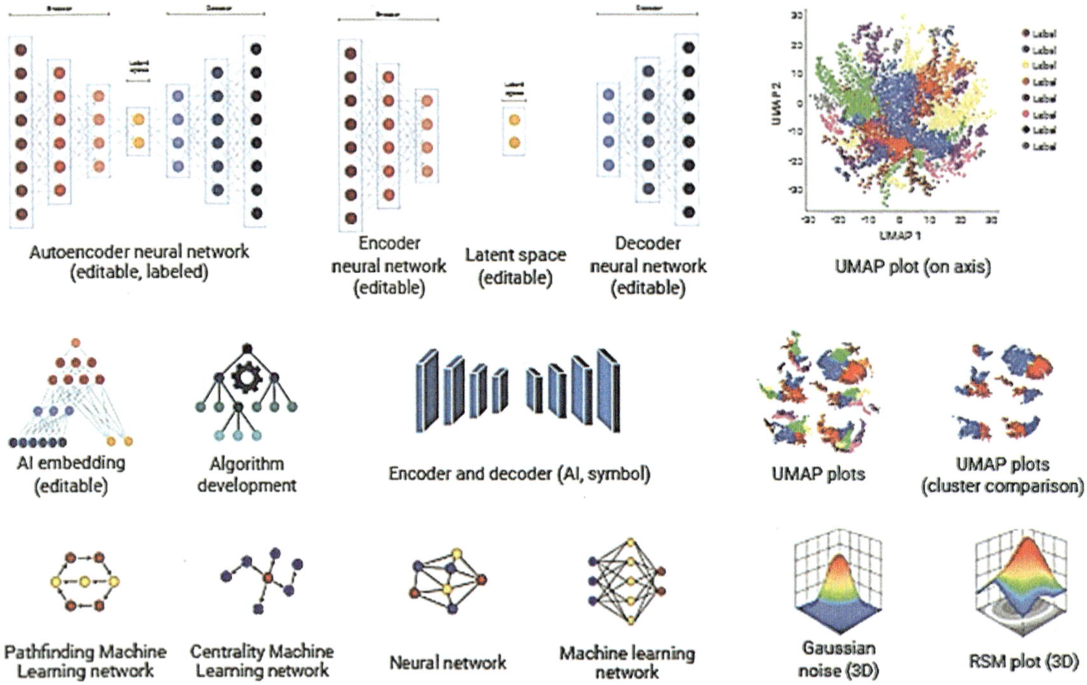

Figure 3.2 Mechanism of data analysis systems in aquaculture.

Furthermore, the use of AI in automating data analytics supports early warning systems, risk assessments, and resource allocation in aquafarms. These technologies also enable precision aquaculture, where individualized monitoring and feeding result in improved growth performance and feed conversion ratios. Data analytics platforms, integrated with cloud computing, allow stakeholders to access insights remotely, enabling data-driven decision-making across the supply chain. This digital transformation is particularly crucial in meeting global food demands, ensuring traceability, and supporting regulatory compliance. As AI models continue to evolve, incorporating techniques such as reinforcement learning and edge computing, the aquaculture sector is poised to benefit from more autonomous, scalable, and cost-effective solutions (Giwa et al., 2022). Thus the synergy between AI and data analytics marks a pivotal shift toward smart aquaculture, promoting environmental sustainability, economic efficiency, and food security.

Blockchain and traceability in aquaculture

The integration of blockchain technology into aquaculture presents a transformative opportunity to enhance traceability, transparency, and trust across the supply chain. In a sector often challenged by issues such as food fraud, antibiotic misuse, and lack of provenance, blockchain offers an immutable, decentralized ledger system that securely records every stage of the aquaculture process from hatchery to harvest, processing, distribution, and retail. Each transaction or event, such as feed intake, health monitoring, antibiotic use, water quality assessments, harvesting, and logistics, is time-stamped and permanently stored on a blockchain network. This enables stakeholders, including producers, regulators, retailers, and consumers, to access verified, real-time data, ensuring that aquaculture products meet safety and sustainability standards (Fig. 3.3). For example, by scanning a QR code on a product package, consumers instantly trace the fish's origin, farming practices, and transportation history. Moreover, blockchain facilitates compliance with international food safety regulations such as those outlined by the Food and Agriculture Organization (FAO) and the

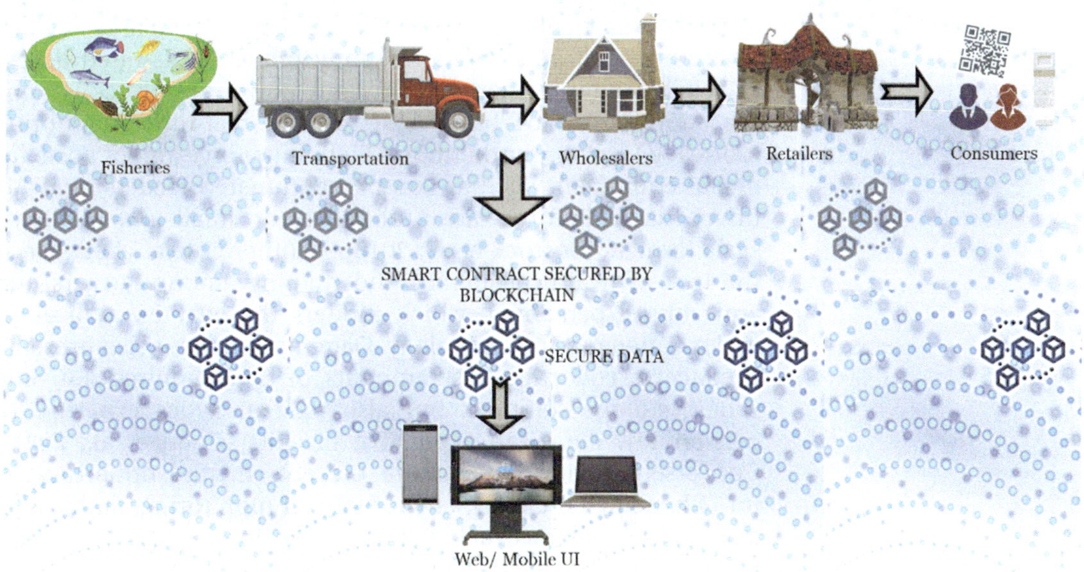

Figure 3.3 Blockchain for aquaculture and fisheries.

European Union, thereby enhancing market access for producers in developing regions. When integrated with other digital technologies, such as IoT devices and AI-driven analytics, blockchain becomes even more powerful, enabling automated data collection and anomaly detection to prevent fraud and ensure quality. Pilot projects in countries such as Norway, Vietnam, and India have already demonstrated blockchain's potential to reduce transaction costs, minimize data tampering, and increase consumer confidence (Galvez et al., 2018; Tian, 2017). As aquaculture continues to expand to meet global seafood demand, the role of blockchain in ensuring sustainable and transparent operations will become increasingly vital. Thus, embracing blockchain-based traceability systems drives innovation, efficiency, and resilience in smart aquaculture ecosystems.

Challenges and opportunities in digital aquaculture

The integration of digital technologies into aquaculture, referred to as digital aquaculture, has emerged as a transformative approach to enhance productivity, sustainability, and traceability in aquatic food production. Despite its promising outlook, the implementation of digital aquaculture faces significant challenges. One of the primary obstacles is the lack of digital infrastructure, particularly in remote coastal and inland regions where aquaculture is predominantly practiced. Limited access to reliable internet connectivity and power supply constrains the adoption of IoT-based systems and real-time monitoring tools (FAO, 2020). Furthermore, high initial investment costs for digital equipment such as sensors, automated feeders, and AI-driven data platforms are prohibitive for small and medium-scale farmers (Bostock et al., 2016). Another critical challenge is the lack of technical literacy and training among farmers and aquaculture personnel, which hinders the effective utilization and maintenance of technology. Data management also presents a concern; collecting vast amounts of data requires not only robust storage solutions but also data analytics expertise to convert raw data into actionable insights (Chen et al., 2022). Conversely, digital aquaculture presents myriad opportunities for enhancing efficiency, environmental monitoring, and profitability. Technologies such as remote sensing, underwater imaging, AI-powered health diagnostics, and blockchain traceability systems enable precision aquaculture, allowing farmers to monitor water quality parameters, detect diseases at early stages, and optimize feeding practices, thereby reducing waste and improving fish health (Misimi et al., 2020). Digital platforms facilitate better decision-making and risk management, which is critical in an industry vulnerable to climate variability and disease outbreaks. Moreover, smart aquaculture fosters transparency and consumer trust, as blockchain and traceability systems ensure end-to-end tracking of aquatic products from farm to fork (Tian, 2016). With increasing global demand for sustainable seafood, there is a strong incentive for governments, startups, and researchers to collaborate and develop scalable, low-cost digital solutions tailored to local needs. In summary, while digital aquaculture faces infrastructural, financial, and skill-related challenges, it also presents unprecedented opportunities for innovation, sustainability, and global competitiveness in aquaculture.

Future perspectives and conclusion

The integration of smart technologies and digital solutions into aquaculture represents a transformative step toward sustainable and resilient food production systems. As global demand for aquatic products continues to rise, driven by population growth, dietary shifts, and economic development, traditional aquaculture practices are increasingly strained by issues such as environmental degradation, disease outbreaks, resource limitations, and the need for traceability. In this context, the application of AI, the IoT, big data analytics, remote sensing, robotics, and blockchain technologies plays a vital role in enhancing productivity, improving resource efficiency, and ensuring the health and welfare of cultured species. In the future, smart aquaculture systems are expected to evolve into fully integrated, autonomous platforms capable of real-time decision-making, predictive modeling, and adaptive management. Precision feeding, automated water quality monitoring, and AI-based health diagnostics will become the norm, reducing labor costs while maximizing yield and sustainability.

Furthermore, the development of user-friendly interfaces and cloud-based data-sharing platforms will facilitate broader adoption among small and medium-scale farmers, bridging the digital divide in aquaculture. However, these advancements also present challenges, including the need for substantial investment, technical expertise, and regulatory frameworks, as well as ethical considerations related to data privacy and environmental impact. Collaborative efforts among stakeholders, governments, academia, industry, and local communities will be essential to overcome these barriers and ensure that the benefits of digitalization are equitably distributed. In conclusion, the future of aquaculture lies in embracing smart and digital technologies not merely as tools for modernization, but as essential components of a holistic approach to responsible and efficient aquatic food production. The transition to smart aquaculture will not only support global food security but also foster innovation, economic growth, and environmental stewardship in the years to come.

References

Ahmed, A., Shah, M. A., & Wahid, A. (2020). IoT-based smart aquaculture: Towards real-time monitoring and efficient management of aquafarms. *Journal of Network and Computer Applications, 156,* Article 102570. https://doi.org/10.1016/j.jnca.2020.102570

Bhattacharyya, R., Kumar, P., & Dutta, R. (2021). Internet of things (IoT) in aquaculture: A review on current trends and future prospects. *Aquacultural Engineering, 94,* Article 102164. https://doi.org/10.1016/j.aquaeng.2021.102164

Bostock, J., Lane, A., Hough, C., & Yamamoto, K. (2016). An assessment of the economic contribution of EU aquaculture production and the influence of policies for its sustainable development. *Aquaculture International, 24*(3), 699–733.

Bostock, J., McAndrew, B., Richards, R., Jauncey, K., Telfer, T., Lorenzen, K., ... Corner, R. (2010). Aquaculture: Global status and trends. *Philosophical Transactions of the Royal Society B: Biological Sciences, 365*(1554), 2897–2912.

Chaudhary, N., Yadav, A., & Prasad, R. (2022). IoT in aquaculture: A smart system for monitoring water quality. *Environmental Monitoring and Assessment, 194*(5), 1–12.

Chen, X., Zou, D., Xie, H., Cheng, G., & Liu, C. (2022). Two decades of artificial intelligence in education. *Educational Technology & Society, 25*(1), 28–47.

FAO (Food and Agriculture Organization of the United Nations). (2020). *Yearbook of International Environmental Law, 31*(1), 326–340.

FAO. (2020). *The state of world fisheries and aquaculture 2020. Sustainability in action.* Food and Agriculture Organization of the United Nations.

FAO. (2022). *Harnessing digital technology for fisheries and aquaculture.* FAO Technical Report.

Føre, M., Frank, K., Norton, T., Svendsen, E., Alfredsen, J. A., Dempster, T., ... Berckmans, D. (2018). Precision fish farming: A new framework to improve production in aquaculture. *Biosystems Engineering, 173,* 176–193.

Galvez, J. F., Mejuto, J. C., & Simal-Gandara, J. (2018). Future challenges on the use of blockchain for food traceability analysis. *TrAC Trends in Analytical Chemistry, 107,* 222–232.

Giwa, T., Akbari, M., & Kumar, A. (2022). Assessment of the greenhouse gas emission footprint of a biorefinery over its life cycle. *Energy Conversion and Management, 271,* 116327.

Hvas, M., Folkedal, O., & Oppedal, F. (2020). Fish welfare in offshore aquaculture systems: A review of current knowledge. *Aquaculture, 528,* Article 735575.

Leng, K., Bi, Y., Jing, L., Fu, H. C., & Van Nieuwenhuyse, I. (2018). Research on agricultural supply chain system with double chain architecture based on blockchain technology. *Future Generation Computer Systems, 86*, 641–649.

Lezzi, M., Lazoi, M., & Corallo, A. (2018). Cybersecurity for Industry 4.0 in the current literature: A reference framework. *Computers in Industry, 103*, 97–110.

Li, X., Zhang, Y., & Wang, H. (2021). AI-driven precision aquaculture: Prospects and challenges. *Computers and Electronics in Agriculture, 185*, Article 106139.

Marr, B. (2020). *Tech trends in practice: The 25 technologies that are driving the 4th industrial revolution.* Wiley.

Misimi, F., Kajtna, T., Misimi, S., & Kapus, J. (2020). Development and validity of the fear of water assessment questionnaire. *Frontiers in Psychology, 11*, 969.

Mohanraj, I., Ashokumar, K., & Naren, M. (2020). Smart aquaculture monitoring system using drones and IoT. *International Journal of Advanced Science and Technology, 29*(4), 10228–10235.

Raghavan, M., Barocas, S., Kleinberg, J., & Levy, K. (2020). Mitigating bias in algorithmic hiring: Evaluating claims and practices. In *Proceedings of the 2020 conference on fairness, accountability, and transparency* (pp. 469–481).

Ramasamy, H., & Vigneswaran, S. (2020). Applications of IoT in smart aquaculture: A review. *Aquaculture Research, 51*(7), 2711–2725. https://doi.org/10.1111/are.14612

Rojas-Torres, J. A., Aguilar, F. A., Rojas, A., & Montoya, L. D. (2022). Artificial intelligence in aquaculture: Applications and future directions. *Aquacultural Engineering, 96*, Article 102218.

Skøien, K. R., Grøtan, T. O., & Foss, A. (2019). Underwater robotic systems for aquaculture: Current status and future perspectives. *Aquaculture International, 27*(4), 1241–1260.

Tian, F. (2016). An agri-food supply chain traceability system for China based on RFID and blockchain technology. In *2016 13th International Conference on Service Systems and Service Management (ICSSSM)* (pp. 1–6).

Tian, Y. (2017). An analytical formula of population gradient for two-layered relu network and its applications in convergence and critical point analysis. In *International conference on machine learning* (pp. 3404–3413). PMLR.

Tsolakis, N., Harris, I., Srai, J. S., & Kumar, M. (2021). Blockchain in agri-food chains: Enabling transparency and efficiency in smart aquaculture. *Production Planning & Control, 32*(1), 55–71.

Verdouw, C. N., Wolfert, J., Beulens, A. J. M., & Rialland, A. (2016). Virtualization of food supply chains with the internet of things. *Journal of Food Engineering, 176*, 128–136.

Xiao, Y., Liu, Y., & Li, X. (2020). Remote sensing applications in aquaculture: A review. *Aquaculture, 526*, Article 735357.

Zhao, J., Bai, J., & Zhou, Y. (2019). Development of smart aquaculture system using wireless sensor network. *Computers and Electronics in Agriculture, 162*, 31–41. https://doi.org/10.1016/j.compag.2019.03.031

Zhao, Y., Wang, X., Liu, Y., Zhang, B., & Xu, C. (2021). Application of internet of things (IoT) for smart aquaculture. *Journal of Sensors, 2021.*

Zhou, J., Wang, J., & Chen, X. (2021). Real-time monitoring and predictive modelling in aquaculture: Integration of IoT and remote sensing technologies. *Sensors, 21*(6), 2131. https://doi.org/10.3390/s21062131

4

Introducing artificial intelligence toward aquaculture

Introduction

Aquaculture is the practice of farming aquatic organisms such as fish, crustaceans, mollusks, and aquatic plants. This sector emerges as a milestone of global food security, economic growth, and sustainable livelihoods. Due to overfishing, habitat degradation, and climate change, wild fish stocks declines which paves way for the emergence of aquaculture as an effective replacement for the rising demand for seafood. Aquaculture contributes more than 50% of the global seafood production says the Food and Agriculture Organization (FAO). In the future, this sector will play an important role in meeting the growing demand (FAO, 2022). In spite of this growth, this sector continues to resist several challenges such as disease outbreaks, ineffective feed management, optimizing water quality, and environmental pollution. These problems indicate that it is necessary to introduce advanced technologies that offer precise, predictive, and sustainable practices to aquaculture. Artificial intelligence (AI), machine learning (ML), deep learning (DL), computer vision, and data analytics has set an importance across multiple sectors such as healthcare, finance, agriculture, and transportation. For the past few years, aquaculture has started to utilize AI resources to overcome the major challenges. AI also provides the potential to analyze large datasets, complex patterns, and make data-based decisions to enhance the efficiency of operation, manage the resources and productivity of aquaculture (Goodfellow et al., 2016; LeCun et al., 2015). For example, to monitor fish behavior, estimate biomass and assess feeding efficiency, computer vision algorithms are used, whereas ML models are highly utilized to predict disease outbreaks and for optimizing water quality parameters (Ahmad et al., 2021). Among the advantages of AI in aquaculture, the ability to connect diverse data sources such as sensor data, environmental parameters, feeding records, and genomic information in order to develop a unique analytical

Aquaculture with Artificial Intelligence. https://doi.org/10.1016/B978-0-443-33704-8.00013-6

framework is a crucial one. This integration makes the establishment of real-time monitoring and automated decision-making in smart aquaculture systems. Such wise systems improves the accuracy of day-to-day operations along with the sustainability of aquaculture which leads to reduction in resource wastage (Gonzalez et al., 2019). For instance, water quality depletion and disease risks are forecasted by predictive algorithms which enables interventions that prevents aqua farms from mortalities and economic losses.

The AI utilization in aquaculture bring into line with wider industry models and agricultural accuracy agriculture, where mechanization, Internet of Things (IoT), big data, and cloud computing unite to redesign outdated industries. Accurate aquaculture enabled by AI allows the tweaked management of aquatic systems by endlessly regulating parameters such as oxygen levels, pH, ammonia levels and feed distribution based on present data analysis (Rani & Sundaram, 2022). These inventions lead to sophisticated yields, improved animal well-being, and condensed dependence on antibiotics and chemicals, conducive to eco-friendly and communally accountable aquaculture practices.

Though, the acceptance of AI in aquaculture is not deprived of obstacles. Technical restrictions, high early venture, data confidentiality concerns, lack of calibration, and inadequate skill in AI among aquaculture experts persist as important blockades. Besides, many small and medium-scale farmers in emerging countries are deficient in essential substructure and cardinal knowledge to incorporate such skills efficiently. Bridging these gaps necessitates interdisciplinary partnerships among AI professionals, marine biologists, aquaculture engineers, legislators, and investors. Training and capacity building programs, combined with management inducements and public–private collaborations, are vital to communalize admittance of AI-driven solutions. This section sets the basis for discovering numerous AI claims in aquaculture, with prominence on technical contexts, real-world situation education, and predictions. As we investigate profoundly, it becomes obvious that accepting AI is not simply a choice but an inevitability for the upcoming flexibility and development of global aquaculture.

Evolution and growth of AI technologies in aquaculture

The progress of AI skills in aquaculture has been equally reformative and advanced, determined by the crucial necessity to improve efficiency, sustainability, and decision-making in aquatic farming systems. Primarily, aquaculture depends severely on labor-intensive and outdated approaches for nursing water quality, fish health, feeding plans, and environmental circumstances. Though over the past 2 decades, progressions in computational models, ML, computer vision, and the IoT have transformed these methods (Fig. 4.1). The incorporation of AI began with the growth of skilled systems in the early 2000s, which were intended to offer rule-based choice backing for aquaculture management (Ayyappan et al., 2007) and with the initiation of ML algorithms, researchers and trade investors began arranging analytical replicas for illness outburst predicting, ideal feeding and growth rate assessment, thereby diminishing feed waste and capitalizing yield (Martins et al., 2011). Computer vision methods, authorized by DL, have additional enablement of immediate monitoring of fish behavior, biomass estimation, and mechanical species identification, crucially dropping the labor-

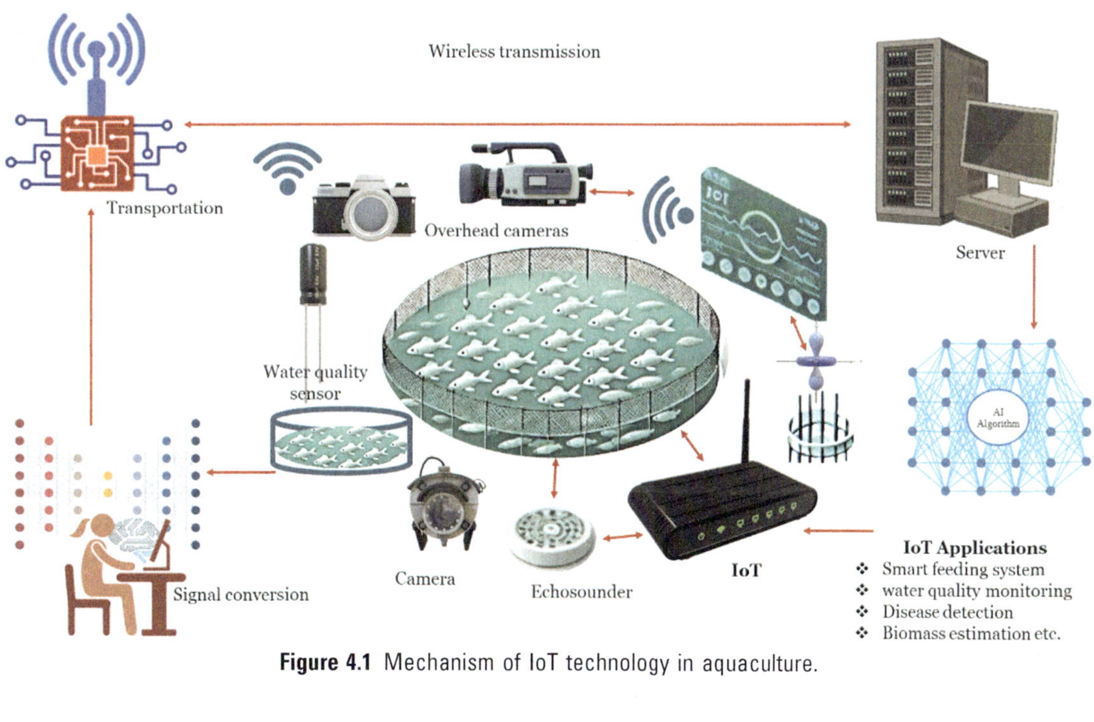

Figure 4.1 Mechanism of IoT technology in aquaculture.

intensive nature of aquaculture operations (Qin et al., 2020). AI-powered water quality devices and IoT stages now let unremitting scrutiny of physicochemical parameters, transferring initial signals around unfavorable variations and enhancing biosecurity procedures (Misra et al., 2022). These clever schemes not only enable improved, healthier farm management but also guarantee environmental amenability and traceability. In addition, current inventions in AI-driven automations are used for programmed feeding, tank cleaning, and underwater investigation, enhancing operational efficiency in inland and marine aquaculture environments. The enduring union of AI with big data analytics and computation is anticipated to socialize these skills further, enabling them reachable to small and medium-scale farmers. This progressing ecosystem underlines a pattern change from sensitive to active aquaculture management. As the global request for aquatic protein rises, the deliberate disposition of AI technologies has made it possible to provision strong, ascendable, and environmentally friendly aquaculture practices worldwide (FAO, 2022; Føre et al., 2018).

Key AI technologies used in aquaculture

AI has arisen as a reformative tool in contemporary aquaculture, offering answers to improve yield, sustainability, and conversant decision-making. One of the most largely used AI technologies in aquaculture is ML, especially managed and unmanaged learning tools, which allow prognostic analytics for development rates, feeding agendas, and disease outbursts (Li, Xu et al., 2020) (Fig. 4.2). ML procedures analyze huge datasets obtained from sensor devices and cameras, recognizing outlines that help enhance farming methods. Deep learning, a subdivision of ML, is highly active in image-appreciating tasks to mechanize fish species identification, evaluate biomass, and display health status via visual cues (Ren et al., 2021). Computer vision models furnished with submerged cameras and convolutional neural networks (CNNs) notice irregular fish behavior and structural changes revealing stress or disease, enabling timely interferences (Chen et al., 2019). Furthermore, IoT devices combined with AI perform a vital role in present environmental monitoring. These strategies endlessly gather data on water quality parameters, including temperature, pH, dissolved oxygen, and ammonia levels. AI models then progress this data to deliver actionable perceptions, empowering farmers to uphold ideal aquaculture

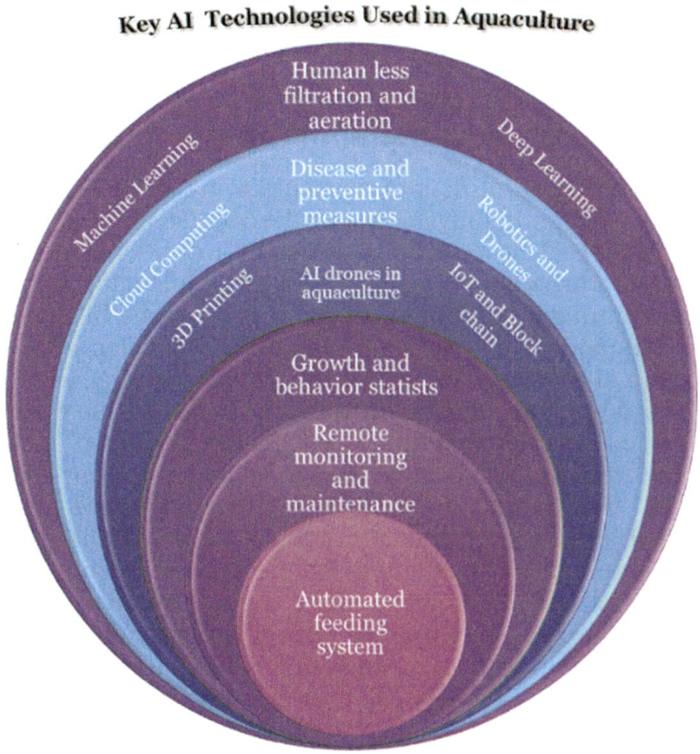

Figure 4.2 Advanced AI technologies used in aquaculture and their uses.

settings (Zhang, Liu, et al., 2020). Reinforcement learning, an alternative AI method, has been discovered for self-governing feeding models, which learn ideal feeding strategies based on fish behavior and need to eat, thereby dropping feed waste and increasing feed conversion ratios (Lee & Kim, 2022). Natural language processing (NLP) is also being merged into shrewd aquaculture administration systems, allowing instinctive human-machine edges and streamlining data interpretation for farmers with less technical knowledge. Also, extrapolative analytics and choice care systems powered by AI aid in predicting market weights, weather effects, and disease epidemics. These tools help calculate arrangement and risk supervision in aquaculture initiatives. The amalgamation of AI with geographic information systems (GIS) allows three-dimensional scrutiny for place assortment, enabling the documentation of optimal locations for aquaculture expansion with negligible ecological impact (Rani et al., 2023). In total, the interaction of AI with harmonizing machineries such as robotics, remote sensing, and cloud computing is transforming aquaculture by empowering meticulous farming, improving operational efficacy, and warranting sustainability.

Applications of AI in aquaculture operations

The incorporation of AI into aquaculture has escorted in a reformative era, importantly increasing the competence, sustainability, and cost-effectiveness of aquaculture operations. AI applications in aquaculture have several realms, including water quality monitoring, disease prediction, feeding systems, behavior observations, and yield forecasting. One of the key potential applications of AI is in instantaneous water quality checking, where sensor-based systems, along with ML algorithms, evaluate parameters such as temperature, pH, dissolved oxygen, and ammonia levels to uphold ideal environmental situations and protect fish from mortality (Gogoi et al., 2019). Predictive analytics, aided by AI, recognize outlines in ecological data to estimate dangerous shifts, thus letting for preventive measures. In the domain of health administration, computer vision and DL techniques have been employed to identify initial symptoms of diseases in fish populations by analyzing photographs or video graphs for irregularities in behavior, color, or body morphology (Zhang, Wang, et al., 2020). Similarly, AI-based mechanized feeding structures improve schedules for feeding and amounts based on fish size, appetite, and environmental conditions, thereby minimizing the feed waste generated and increasing growth rates (Kim et al., 2021). Behavioral analysis using AI algorithms aids in tracking fish activity, grouping patterns, and stress pointers, which are important for safeguarding animal well-being and productivity. Moreover, AI-empowered decision care systems support farm managers in deliberate preparation by incorporating past and present data to predict yield, assess economic performance, and rationalize supply chain logistics (Ghamisi et al., 2022). Drones and automated underwater vehicles, assisted by AI algorithms, are progressively used for monitoring and stock evaluation in large-scale aquaculture arrangements. These technical involvements decrease the labor cost and man-made error, along with the contribution to sustainable aquaculture through endorsing responsible resource management and dropping environmental impacts. As AI endures to change, its merging with IoT, big data analytics, and robotics promises even more wise and self-directed aquaculture procedures in the future.

Case studies and success stories

The reformative effect of AI on aquaculture has been established through various worldwide case studies and victory stories, highlighting its potential to enhance yield, sustainability, and reserve material competency. For example, *eFishery*, an Indonesian aqua-tech industry, transformed outdated aquaculture methods by implementing smart feeding systems that utilize AI -empowered sensors to observe fish behavior and idealize feed circulation (Fig. 4.3). This invention results in a 21% hike in feed efficacy and a 28% rise in profit outlines for farmers (Setiawan et al., 2021). Likewise, *Aquaai*, a U.S.-based firm, installed AI-supported mechanized fish for real-time observation of water quality indicators, fish well-being, and underwater behavior, permitting proactive disease identification and moderation, thus crucially decreasing mortality counts (Lee & Kim, 2020). In Norway, the implementation of ML algorithms into salmon farming activities by industries such as *Cermaq* and *BioSort* has facilitated computerized lice recognition and biomass estimation

Figure 4.3 Smart feeding systems that use AI-powered sensors to monitor fish behavior and optimize feed distribution.

using underwater photographic devices and computer vision, consequently allowing for more specific treatments and increased yield prediction (Abolofia et al., 2020). Indian enterprises such as ICAR-CIFA's pilot project on AI-assisted water quality prediction and choice systems have proven results in small—scale inland aquafarms by offering budget-friendly solutions specific to regional necessity (Rajesh et al., 2022). These illustrations together highlight how AI is improving operational productivity and success, simultaneously encouraging sustainability via data-aided decision-making. The acceptance of AI in aquaculture is thus not a theoretical invention but a demonstrated approach with quantifiable consequences in varied aquaculture systems globally.

Challenges and limitations of AI adoption in aquaculture

In spite of the growing interest in incorporating AI into aquaculture, numerous encounters and boundaries delay its extensive acceptance and efficiency. One of the most important obstacles is the lack of good and consistent data. AI replicas require great capacities of precise and well-explained datasets for training and authentication. Though aquaculture systems often cause unpredictable or unfinished data due to environmental inconsistency, varied species, and changes in farming practices (Li, Wang, et al., 2020). Furthermore, structural restrictions such as the nonappearance of dependable internet connectivity in rural aquaculture regions limit real-time data collection and cloud-driven AI processing. High early costs for applying AI-based solutions, including sensing devices, IoT systems, and analytical software, are some obstacles, especially for small and medium-scale farmers with inadequate capital investment (Ahmed et al., 2021). There is also an information gap among aquaculture experts who may be deficient in the technical skills to work with AI systems, understand results, or solve technical failures. This requires wide training programs and interdisciplinary cooperation between aquaculture practitioners and data scientists. In addition, regulatory concerns such as data privacy, ownership, and capable dislocation of physical labor must be considered to confirm reasonable distribution of expertise (Rajabi et al., 2022). The other restriction is the background flexibility of AI models, as many of the systems are practiced in closed measurable environments, they may not accomplish well beneath varied and active aquatic circumstances, which necessitates nonstop prototypical informing and localization. Lately, governing and policy frameworks for AI use in aquaculture are yet weak in several countries, which results in doubt regarding application values and data dominance. Overcoming these obstacles is important while unleashing the full capability of AI in improving aquaculture efficiency, sustainability, and flexibility.

Future perspectives and strategic recommendations

The mixing of AI into aquaculture holds reformative capacity to transfigure the field by idealizing yield, improving sustainability, and enhancing decision-making processes. Looking forward, the future of AI in aquaculture is composed of shifting toward more mechanized and adaptive models that pull real-time data from sensing devices, drone cameras, and satellite photographic machines to achieve feeding commands, monitor water quality, and identify disease with accuracy. The expansion of predictive systems using ML and DL will further permit early signs of warnings for epidemic outbursts and environmental pollutants, thus decreasing mortality and loss. To utilize all the advantages of AI, planned reserves must be focused in the direction of research and growth, structural upgradation, and multidisciplinary partnerships between data scientists, aquaculture personnel, and policymakers. Educating programs and capacity building ideas are important to resolve the existing gap and support investors to functionalize AI-supported systems efficiently.

Moreover, it is overbearing to implement homogeneous procedures for data obtaining, sharing, and privacy to enable large-scale applications of AI techniques. Regulatory issues such as open algorithmic decision-making, social and economic allegations of mechanization must be resolved to make sure the collective growth and reasonable access to AI tools. Democratic parties and funding partners should offer support through pilot projects and public–private associations that can bring the large-scale and acceptability of AI utilization in various aquaculture conditions, from small-scale farms to large marine systems. To conclude, tactical incorporation of AI techniques in aqua industries directed by advanced thinking policies, sustainable methods, and collective inventions improves efficacy and environmental support, along with long-term assurance and affordability of the aquaculture industry in accordance with the global obstacles like climate change and food security.

Conclusion

AI techniques provide great capacity in augmenting feeding methods, identifying disease spread, maintaining water quality, and mechanizing farm administration systems. While binding ML algorithms with computer vision and analytics, aquaculture functions crucially increase its power and reduce the operational cost simultaneously with less environmental effects. The acceptability of AI in immediate choice-making drives farmers with informed decisions, accurate predictions, and safe aquaculture approaches. Since the global need for seafood rises continuously, the incorporation of AI offers a proven pathway to overcome the yield targets and maintain environmental balance. Moreover, the fruitful acceptance of AI needs multidisciplinary association, vigorous inventions, and training steps to make sure convenience and scalability. Finally, the combination of AI and aquaculture has the weapon to carve a more capable, data-driven, and technically superior future of aquatic farming.

References

Abolofia, J., Asche, F., & Wilen, J. E. (2020). The cost of lice: Quantifying the impact of parasitic sea lice on farmed salmon. *Marine Resource Economics, 35*(2), 125–138.

Ahmad, Z., Rahim, S., Zubair, M., & Abdul-Ghafar, J. (2021). Artificial intelligence (AI) in medicine, current applications and future role with special emphasis on its potential and promise in pathology: present and future impact, obstacles including costs and acceptance among pathologists, practical and philosophical considerations. A comprehensive review. *Diagnostic Pathology, 16*(1), 24.

Ahmed, N., Thompson, S., & Glaser, M. (2021). Artificial intelligence and big data in aquaculture: A review. *Aquaculture Reports, 20,* Article 100700.

Ayyappan, S., Jena, J. K., Gopalakrishnan, A., & Pandey, A. K. (2007). *Handbook of fisheries and aquaculture.* Indian Council of Agricultural Research (ICAR).

Chen, Z., Li, J., & Xu, X. (2019). Application of computer vision in aquaculture. *Aquaculture Reports, 14,* 100–108.

Føre, M., Frank, K., Norton, T., Svendsen, E., Alfredsen, J. A., Dempster, T., & Berckmans, D. (2018). Precision fish farming: A new framework to improve production in aquaculture. *Biosystems Engineering, 173,* 176–193.

FAO. (2022). *The state of world fisheries and aquaculture 2022.* Food and Agriculture Organization of the United Nations.

Ghamisi, P., Yokoya, N., Li, J., Liu, Y., & Plaza, A. (2022). Advanced artificial intelligence techniques in aquaculture systems: Emerging trends and future prospects. *Trends in Biotechnology, 40*(4), 486–498.

Gogoi, B. J., Baruah, D. C., & Hazarika, M. K. (2019). Application of IoT and AI in aquaculture: A review. *Journal of Fisheries and Life Sciences, 4*(2), 45–52.

Gonzalez, M. F., Capman, J. F., Oswald, F. L., Theys, E. R., & Tomczak, D. L. (2019). "Where's the IO?" Artificial intelligence and machine learning in talent management systems. *Personnel Assessment and Decisions, 5*(3), 5.

Goodfellow, I., Bengio, Y., Courville, A., & Bengio, Y. (2016). *Deep learning.* Cambridge: MIT Press.
Kim, S. K., Kim, J. Y., & Kim, K. H. (2021). Smart aquaculture feeding systems using AI and computer vision technologies. *Computers and Electronics in Agriculture, 188,* Article 106348.
LeCun, Y., Bengio, Y., & Hinton, G. (2015). Deep learning. *Nature, 521*(7553), 436–444.
Lee, Y., & Kim, J. (2020). Smart monitoring system using AI-based robotic fish for aquaculture. *Journal of Aquaculture Engineering and Technology, 15*(3), 109–118.
Lee, H., & Kim, Y. (2022). Smart feeding systems using reinforcement learning in aquaculture. *Computers and Electronics in Agriculture, 193,* Article 106653.
Li, X., Wang, Y., & Zhao, H. (2020). Machine learning applications in aquaculture: A review. *Aquaculture International, 28*(5), 1533–1550.
Li, X., Xu, J., Wang, Y., & Wang, W. (2020). Smart aquaculture: Opportunities and challenges. *Sensors, 20*(19), 5595.
Martins, C. I., et al. (2011). New developments in recirculating aquaculture systems in Europe: A perspective on environmental sustainability. *Aquacultural Engineering, 43*(3), 83–93.
Misra, S., Roy, A., & Obaidat, M. S. (2022). Smart aquaculture: A review on recent advances, challenges, and future directions. *IEEE Internet of Things Journal, 9*(1), 763–781.
Qin, J., Wang, D., Zhang, Y., & Zhou, C. (2020). Deep learning for underwater image enhancement and object detection: A review. *IEEE Access, 8,* 123456–123471.
Rajabi, S., Ranjbar, A., & Arasteh, A. (2022). Ethical considerations in AI-based aquaculture systems. *Journal of Aquaculture Technology, 14*(1), 45–52.
Rajesh, K., Sahoo, L., & Nanda, P. (2022). Development of AI-based water quality prediction model for inland aquaculture. *Indian Journal of Fisheries Science and Technology, 8*(1), 45–52.
Rani, R., Kumar, V., & Singh, M. (2023). GIS and AI integration for aquaculture site selection. *Environmental Modelling & Software, 164,* Article 105543.
Rani, V. S., & Sundaram, N. (2022). Collaborative social media marketing in small scale business using artificial intelligence. *Electrochemical Society Transactions, 107*(1), 5175–5182.
Ren, L., et al. (2021). Deep learning for fish detection and species classification in aquaculture. *Sensors, 21*(8), 2792.
Setiawan, H., Prasetyo, A. A., & Nugroho, H. (2021). Impact assessment of AI-integrated feed automation on smallholder aquaculture productivity: The case of eFishery in Indonesia. *Aquaculture Reports, 20,* Article 100676.
Zhang, Q., Liu, H., & Wu, Z. (2020). IoT-based water quality monitoring and AI analysis in aquaculture. *Journal of Cleaner Production, 276,* Article 124208.
Zhang, D., Wang, H., & Li, Z. (2020). Deep learning-based fish disease recognition using underwater images. *Aquaculture, 523,* Article 735197.

5

Learning algorithms of AI and ML for aquaculture

Introduction

Aquaculture, or the cultivation of aquatic animals, such as fish, shellfish, and marine plants, has been a mainstay of global food security for centuries. The global population is expected to grow to more than nine billion by 2050 as the production of healthy and sustainable protein expands. Classical capture fisheries are already at or past their biological capacity, rendering aquaculture the rapidly developing sector of the international food production system. In 2022 alone, aquaculture accounted for over 50% of the world's fish consumption, a proportion expected to grow steadily over the coming decades. Its function in reducing hunger, facilitating livelihoods, and stimulating economic growth, especially among coastal and rural populations, is currently understood and accepted by agencies, such as the United Nations Food and Agriculture Organization and the World Bank (Boyd et al., 2022). However, due to its unmatched potential, aquaculture in the present times still suffers from significant operational, environmental, and biological problems that render its sustainability and resilience difficult to attain. Management of diseases remains one of the most insurmountable difficulties in aquaculture. Bacterial, viral, parasitic, and fungal infectious diseases continue to be the primary determinants of financial loss in aquaculture industries worldwide. Often, outbreaks of disease may destroy complete inventories within days, resulting in substantial economic losses and posing a risk to the supply chain. Traditional disease monitoring procedures, which are highly reliant on sample collection and human laboratory analysis, are both time-consuming and labor-intensive, and typically reactive rather than proactive (Martinez-Porchas & Martinez-Cordova, 2012).

Furthermore, the increasing prevalence of antimicrobial resistance, which is exacerbated by the irresponsible use of antibiotics in aquaculture, complicates the effort to manage the

Aquaculture with Artificial Intelligence. https://doi.org/10.1016/B978-0-443-33704-8.00008-2

disease. Yield maximization is yet another critical issue. Maintaining stable and reproducible yields requires tight control over a range of environmental conditions, including water temperature, oxygen levels, pH, and nutrient levels. Deviation from any of these can result in below-optimum growth rates, increased mortality, and reduced profitability. Conventional approaches to farm management, which rely heavily on practical experience and intuition, often prove ineffective in addressing the complex, nonlinear interactions characteristic of aquatic ecosystems (Mohammed et al., 2025).

Resource management is also an added challenge. Water, energy, and feed are primary inputs for aquaculture production, and their wasteful use not only raises operating expenses but also contributes to environmental degradation. Excess nutrient runoff makes water bodies eutrophicate, and inappropriately managed feed creates waste and disrupts the ecosystem. Precision aquaculture solutions are sorely needed that will reduce resource consumption, eliminate waste, and promote environmental stewardship. Against this backdrop, Artificial Intelligence (AI) and Machine Learning (ML) are emerging as game-changing technologies that can be used to address most of aquaculture's long-standing issues. AI, or the potential for machines to carry out tasks that hitherto required human intelligence, and ML, an AI subdiscipline aimed at enabling systems to learn to make decisions based on data and improve over time without being explicitly programmed, present unrivalled potential for boosting decision-making, process automation, and resource optimization in aquaculture (Dauda et al., 2019). By leveraging large volumes of data from bioassays, environmental sensors, video cameras, and smart sensors, ML and AI can detect latent patterns, forecast outcomes, and facilitate preventive interventions. ML algorithms, for example, can be trained on past environmental and health data to forecast disease outbreaks before the onset of clinical symptoms, triggering early preventive interventions. AI-powered deep learning computer vision technology can capture anomalies in form, motion, or behavior in fish, those earliest signs of health issues that tend to go unnoticed by human observers (Niazi, 2023).

In yield optimization, ML algorithms can analyze complex interactions between various environmental factors and measures of growth performance to recommend control strategies tailored to specific species and production systems. Reinforcement learning (RL) models can optimize feeding regimes in real time, modifying feed amounts according to fish appetite and metabolic requirements, thus improving growth efficiency and reducing feed wastage. In addition, AI-based decision support systems can combine multisource data, including from meteorological predictions to market price patterns, to support farm managers in strategic planning and risk management (Zarbakhsh et al., 2025). Resource management will also benefit immensely from the use of AI and ML. Intelligent monitoring systems can detect deviations in water quality parameters and trigger control actions, such as aeration or filter adjustments, in real time. Predictive models of maintenance have the potential to forecast equipment failure before it occurs, thereby avoiding downtime and operational disturbances. At a larger scale, AI-based ecosystem modeling can help design sustainable farm plans, estimate carrying capacity estimates, and reduce ecological footprints. Significantly, AI and ML usage is not limited to high-scale industrial farms. With advancements in low-cost sensors, cloud computing, and open-source ML libraries, even small- and medium-scale aquaculture enterprises can leverage these technologies to enhance their sustainability and productivity (Essamlali et al., 2024).

Fundamentals of AI and ML in aquaculture

AI and ML have become revolutionary technologies in various industries, including agriculture, healthcare, and industrial automation. Aquaculture, one of the most rapidly growing food-producing industries worldwide, is increasingly adopting AI and ML to address traditional challenges, such as disease outbreaks, resource inefficiency, environmental fluctuations, and yield maximization. With increasingly data-intensive aquaculture systems resulting from the proliferation of sensors, computer vision, and remote monitoring technologies, AI and ML present an unprecedented opportunity to leverage these streams of data to create actionable insights and predictive capabilities (Zatsu et al., 2024). Essentially, AI is the simulation of human intelligence processes by machines. In contrast, ML is a subset of AI that enables systems to learn from data without being explicitly programmed. ML models are trained to recognize intricate patterns, learn from past trends, and predict or make decisions with minimal human intervention. The application of ML in aquaculture enables real-time environmental monitoring, behavioral analysis, feeding optimization, disease diagnostics, and practical farm management (Xu et al., 2021). ML techniques are classified into three broad learning paradigms: supervised learning, unsupervised learning, and RL. Supervised learning involves training models from labeled data for which the input–output relationship is known. The method has a wide application in aquaculture, including fish disease categorization, biomass estimation, and water quality estimation. Unsupervised learning, however, operates on unlabeled data to identify hidden patterns or clusterings and is particularly helpful for behavioral clustering of aquatic animals and sensor anomaly detection. RL is a goal-directed learning process where an agent learns to make decisions through trial-and-error interactions with an environment that changes over time. RL is also under investigation for aquaculture to optimize feeding regimes and energy consumption in computerized agriculture (Sarker, 2021b). The application of these learning algorithms in aquaculture entails the ability to handle high, complex, and nonlinear data characteristic of aquatic systems while producing stable solutions that improve over time. AI systems can help facilitate timely intervention by detecting patterns in fish behavior, environmental variation, or equipment performance. By doing so, losses are mitigated, and productivity is improved. Automation using AI and ML also saves labor, maximizes operational efficiency, and conserves resources (Fini et al., 2025).

Supervised learning

Supervised learning is one of the most common applications of ML in aquaculture, where data are learned from labeled examples. In labeled data, input–output pairs are precisely specified, allowing the model to understand the mapping function from inputs (e.g., sensor readings, water conditions) to desired outputs (e.g., species classification, health status). The model learns from training data and uses the learned mapping function to predict unseen, new data (Essamlali et al., 2024).

Principal algorithms in supervised learning are:

1. Decision trees: Easy to interpret and simple models for multiclass or binary classification.
2. Random forests: Ensemble algorithms that minimize overfitting and improve accuracy by voting decision trees.
3. Support vector machines (SVMs): Strong classifiers that identify the best hyperplanes for separating data, particularly when working in high-dimensional space.
4. Artificial neural networks (ANNs): Nonlinear models for learning able to identify complex patterns in biological and environmental data (Sarker, 2021b).

Applications in aquaculture include:

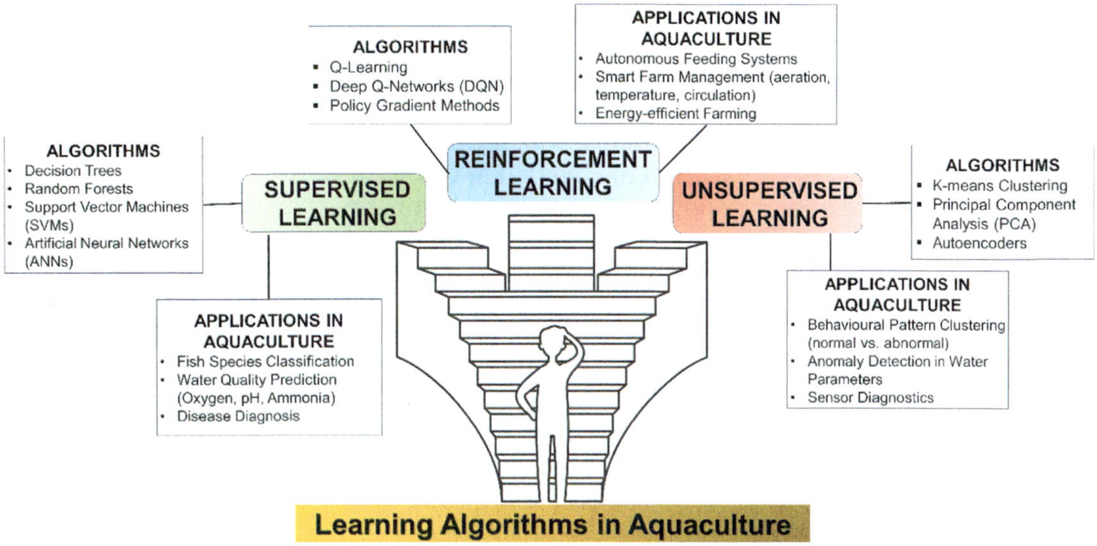

Figure 5.1 Learning algorithms in aquaculture.

1. Species classification of fish: Automatically classifying species based on visual or genetic data for harvesting or sorting.
2. Water quality prediction: Prediction of dissolved oxygen, pH, or ammonia concentration based on sensor readings and environmental conditions.
3. Diagnosis of disease: Detection of pathological epidemics at an early stage by categorization of symptomatic patterns from fish behavior, histology, or imaging data.

These models prove useful when annotated datasets are available, allowing them to predict accurately and provide early intervention for farm operations (Fig. 5.1) (Abdullah et al., 2024).

Unsupervised learning

Where labeled data are not available or limited, unsupervised learning algorithms provide valuable insights by revealing hidden patterns or structures within the data. These models cluster or summarize data automatically without prespecified output labels.

Some popular algorithms are:

1. K-means clustering: Divides data into k-groups based on similarity and can be employed for behavioral categorization.

2. Principal component Analysis: A linear dimensionality reduction approach that yields principal components from data, best adapted to visualize high-dimensional biological data.
3. Autoencoders: Neural networks that reduce data to latent representations, which can be used for anomaly detection and pattern discovery (Liu et al., 2023).

Applications in aquaculture are:

1. Behavioral pattern clustering: Classifying fish movement or feeding behavior into normal and abnormal patterns, which may reflect stress, disease, or aggression.
2. Anomaly detection in water parameters: Identifying outliers in sensor data that may indicate leaks, chemical spills, or system malfunctions.
3. Unsupervised models are advantageous in exploratory analysis, sensor diagnostics, and early-stage model development, where human labeling is impractical or cost-prohibitive (Fig. 5.1) (Shreesha et al., 2023).

Reinforcement learning

RL is a self-learning ML framework in which agents learn to make decisions based on their experience of interacting with an environment. The agent is rewarded or penalized depending on the actions taken and eventually learns an optimal policy for maximizing cumulative rewards (Alkam et al., 2025).

The main algorithms are:

1. Q-learning: An approach based on values that learns what to do best in a state.
2. Deep Q networks: Combines Q-learning with deep neural networks to deal with high-dimensional, complex state spaces.
3. Policy gradient methods: Improve policy function directly by ascending gradients of expected reward (Serrano, 2022).

Aquaculture uses

1. Autonomous feeding systems: RL agents learn to adjust feed quantity and schedule in real time based on sensor feedback (e.g., turbidity, appetite sensors), reducing feed wastage and promoting growth rates.
2. Smart farm management: Adaptive learning agents control aeration, temperature control, and circulation systems by learning the impact of various actions on fish well-being and energy efficiency (Soussi et al., 2024).

The adaptability of RL makes it best suited for real-time control systems in complex aquaculture scenarios, where feedback is lagged and nonlinear (Fig. 5.1).

Deep learning techniques

Deep learning, an ML subdiscipline, leverages multilayer neural networks that can learn end-to-end hierarchical features from raw input data. The models are well-suited to handle massive, high-dimensional, and complex data, such as images, video, and time series data, which are commonly found in aquaculture (Sarker, 2021a).

Some of the best models are:

1. Convolutional neural networks (CNNs): They are best suited for spatial data and are most used in image recognition and object detection.
2. Recurrent neural networks: Suitable for sequence data, such as sensor time series and growth curves.
3. Transfer learning: Refers to retraining a deep learning model that was pretrained on everyday objects (e.g., VGG, ResNet) for fine-tuning relevant aquaculture tasks using a small quantity of training data (Shin et al., 2016).

Use cases in aquaculture:

1. Fish health monitoring through imaging: Detects wounds, infections, or parasites from waterborne images using CNNs.
2. Adaptation to local conditions: Transfer learning enables global models to become more conducive to fine-tuning with local farm data, thereby significantly lowering training costs and time (Zhang et al., 2022).

Deep learning models are very data-hungry, but they offer an unrivaled level of accuracy and feature extraction, which is invaluable in advanced aquaculture analytics and automation.

Application case studies

Incorporating AI and ML technologies into aquaculture operations has ushered in a new revolution for aquaculture fish farming and aquatic resource management. Traditional aquaculture companies have been plagued for years by unexpected environmental conditions, disease outbreaks, inefficient resource utilization, and equipment breakdown, which have impacted yield, operating costs, and ecological sustainability. Recent developments in AI and ML have enabled novel solutions, making it possible to achieve accurate monitoring, prediction, and

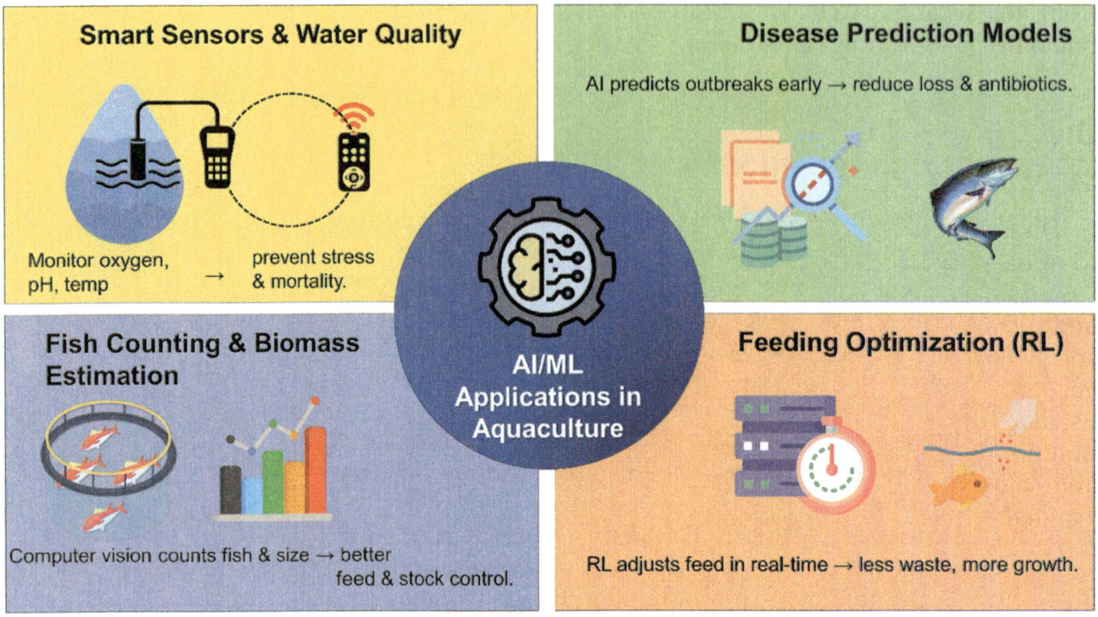

Figure 5.2 Applications of AI/ML in aquaculture.

decision-making functionalities that were previously impossible using traditional approaches. This section discusses key application case studies where AI and ML are transforming aquaculture operations in several aspects (Fig. 5.2) (Ashraf Rather et al., 2024).

Smart sensors and ML-enabled oxygen, pH, and temperature

Monitoring water quality parameters, such as dissolved oxygen, pH, and temperature, is significant in influencing fish health, growth rate, and survival. Minor alterations can trigger stress reactions, lower feed efficiency, or result in mass mortality events. Sophisticated sensor networks coupled with AI-based analytics have been developed to enable real-time, continuous monitoring of these critical parameters. ML models trained on historical water quality data can predict anomalies and initiate autonomous actions, such as activating aerators or adjusting water flow rates. Predictive models not only improve the operational efficiency of farms but also predict ecological risks by maintaining optimal aquatic conditions. Additionally, these intelligent systems reduce workforce requirements, allowing farm administrators to oversee a variety of farms simultaneously from any location, thereby providing improved scalability and reliability (Sáez-Arteaga et al., 2024).

Disease prediction models based on supervised ML

Disease outbreaks remain one of the most significant threats to aquaculture productivity and sustainability. Traditional disease management techniques heavily rely on visual examination and postoutbreak treatment, often resulting in reactive responses and substantial economic losses. Supervised learning algorithms, such as Random Forests, SVMs, and ANNs, have been used to make predictions about disease outbreaks based on multimodal information sources, including water quality indices, anomalous behavior patterns, and early biomarker indications. Trained on large datasets of historical outbreaks, the predictive models recognize very fine-grained patterns and relationships that cannot be detected manually, thereby providing helpful early warning signals and anticipatory health management mechanisms. Early treatment not only lowers mortality rates but also reduces the use of antibiotics and chemical treatments, aligning with sustainable aquaculture methods and customer demands for chemical-free seafood (Irshath et al., 2023).

Fish counting and biomass estimation using computer vision

Precise estimation of fish population and biomass is vital for efficient inventory control, informed feeding practices, and accurate production forecasting. Human counting using classical methods is manual, time-consuming, error-liable, and intrusive in terms of disturbing aquatic life. Deep learning–powered technologies, such as CNN, provide noncontact, real-time solutions for counting and estimating biomass in fish. Submerged sensors and ML algorithms are capable of automatically detecting, counting, and classifying fish, even under turbid or changing water conditions. Furthermore, advanced algorithms can even estimate the size distribution of fish and disease status, capturing entire data for precision agriculture. Such technologies support enhanced decision-making, improve feed management practices, and enable the efficient use of resources by preventing feed wastage (Wägele et al., 2022).

Optimization of automated feeding systems using RL

Feed forms the highest cost of operation in aquaculture, accounting for as much as 60% of the overall expenditure. Both overfeeding and underfeeding are not only wasteful but also impair water quality and enhance disease pressure. RL algorithms provide a dynamic solution by optimizing feeding patterns based on constant feedback derived from fish behavior, environmental parameters, and growth rate. Through simulating thousands of feeding scenarios, RL-based systems can dynamically optimize feeding frequencies and amounts to optimize growth performance and minimize waste. Some deployments combine computer vision systems for real-time monitoring of fish appetite signs to further improve feeding accuracy. This closed-loop optimization approach significantly improves feed conversion ratios (FCRs), operational profitability, and environmental sustainability (Abdullah et al., 2024).

Aquaculture equipment predictive maintenance

Aerator breakdown, pump breakdown, and sensor drift are equipment failures that can devastate farm operations if not caught in their nascent stages. Predictive maintenance, driven by AI-powered anomaly detection and failure prediction, provides a proactive solution for equipment management. By analyzing streams of sensor data and maintenance history, ML algorithms can identify precursors to mechanical wear and tear, schedule preventive maintenance, and automate maintenance tasks. This reduces unplanned downtime, extends equipment lifespan, and lowers operating costs. Moreover, the integration of predictive maintenance into overall farm management systems enables holistic operating optimization, which aligns with the principles of smart and sustainable aquaculture (Ahmed Murtaza et al., 2024).

All these application case studies collectively demonstrate how AI- and ML-related technologies are transforming aquaculture to make it an efficient, sustainable, and resilient industry. As the industry continues to develop further, the increased integration of AI-based innovations will play a crucial role in achieving global food security with reduced ecological costs (Yang et al., 2025).

Data requirements and challenges

The effectiveness of AI and ML technology in aquaculture relies most heavily on access to high-quality, representative data. Data serve as the foundation for any AI/ML workflow on which predictive models are developed, tested, and released. The nature of aquaculture, with its variable environmental conditions, biological variability, and complexity of aquatic systems, means that data play an even more critical role. Without robust, reliable, and contextually informative datasets, even the most sophisticated learning algorithms risk underperformance, biased results, and poor generalization in actual farm environments (Saleh et al., 2024). Good-quality aquaculture data must meet some fundamental standards: accuracy, completeness, relevance, consistency, and timeliness. Adequate data are such that measurements reflect actual environmental and biological conditions, and completeness is such that key parameters, including water quality indicators (e.g., dissolved oxygen, pH, temperature, ammonia concentration), stock biomass, feeding rates, and health measures, are well covered over time. Relevance refers to the appropriateness of collected features for ML tasks, such as fish species identification, disease prediction, or environmental forecasting. Consistency across collection methods diminishes systemic errors, while speed enables the development of real-time monitoring and decision-assistance systems. Despite the intuitive worth of good-quality data, several long-standing challenges limit the applicability of AI and ML in aquaculture applications (Fig. 5.3) (Chandran et al., 2025).

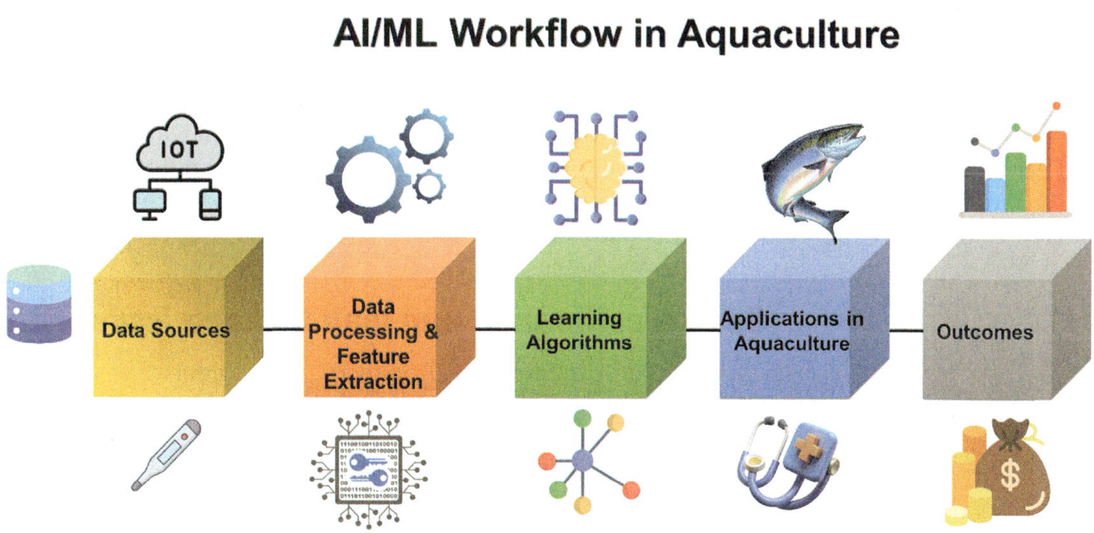

Figure 5.3 AI/ML workflow in aquaculture.

Limited labeled data

One of the main bottlenecks is the lack of labeled datasets, especially those with expert biological knowledge. Supervised learning algorithms require labeled data where input–output pairs need to be defined explicitly. Annotating data in aquaculture, for instance, such as diagnosing diseases in fish from images, behavior classification, or annotating feeding patterns, necessitate domain-specific expertise that is both expensive and time-consuming to obtain. Additionally, the absence of standardized labeling protocols on farms and across regions intensifies heterogeneity, resulting in data heterogeneity that discourages model transferability and stability. This shortage is likely to compel researchers to either use small datasets with a high risk of overfitting or resort to semisupervised and transfer learning methods, which, however, may still be inadequate without sufficient annotated examples (Zhang et al., 2023).

Data variability (seasonal, species-specific)

Aquaculture data are highly variable at both spatial and temporal scales. Monsoon-induced variations in water temperature, salinity, and nutrient concentrations impose nonstationarity on environmental factors. In addition, species-specific variations in physiological, behavioral, and disease susceptibility further complicate model generalization. For example, a trained AI model designed to predict disease onset in Atlantic salmon may not perform reliably when transferred to tilapia or shrimp production systems without extensive retraining or adjustments. In addition, environmental stressors, stocking densities, farm management, and local ecological dynamics all interact in a complex, often nonlinear fashion, making the landscape of modeling highly heterogeneous. Mitigating such heterogeneity requires approaches such as dynamic model updating, domain adaptation, and contextual metadata incorporation during training (Chilton et al., 2021).

Sensor calibration and maintenance issues

The volume of sensor technologies, from underwater cameras and acoustic sensors to IoT-based water quality sensors, has increased the volume of data available in aquaculture exponentially. The reliability of AI and ML solutions, however, depends on the quality of the measurements made by sensors. Sensors in water-based networks are susceptible to biofouling, drifting, and aging, leading to inaccurate measurements and missing data. Maintenance, cleaning, and calibration are essential to maintain data integrity, but they are time-consuming and may be performed irregularly across farms, particularly in limited-resource environments. Natural sources of noise and bias, such as turbidity, sedimentation, and biofilm growth, also affect sensor performance and introduce bias into datasets. Defective or uncalibrated sensors can mislead ML models, leading to incorrect predictions and potentially costly management actions (Biazi & Marques, 2023). The combined effect of all these defects necessitates an active learning in data management for AI-driven aquaculture. Techniques such as active learning to maximize labeling efforts, data augmentation to synthetically enhance datasets, anomaly detection to indicate sensor failure, and extensively validated data validation pipelines are required to offset the constraints of current data environments. Moreover, interdisciplinarity among marine biologists, aquaculture scientists, AI engineers, and sensor experts is necessary to standardize data collection protocols, promote open data initiatives, and enhance dataset quality and availability across the global aquaculture industry (Sriputhorn et al., 2025).

Model optimization and evaluation

Strong model optimization and testing are essential to the efficacy of AI and ML technologies in aquaculture. Since these models hold a central position in decision automation activities, ranging from water quality forecasting to fish health observation, their performance should be rigorously tested using the right metrics and optimized through systematic techniques. A strong evaluation framework ensures that models not only perform well on training data but also generalize well to unseen, real-world situations characteristic of dynamic aquaculture environments (Yang et al., 2025). Model performance is typically assessed using a battery of evaluation metrics. Accuracy, the proportion of correctly predicted instances to the total cases, is a straightforward measure but can be misleading with imbalanced data, a typical case in aquaculture, such as in rare disease detection. Precision and recall provide a more differentiated view: precision verifies the ratio of accurate optimistic predictions to all positive predictions, whereas recall verifies the ratio of correct optimistic predictions among actual positives. The F1 Score, which is the harmonic mean of precision and recall, balances the two and is particularly useful in the situation of an imbalanced class distribution. For regression problems with continuous responses, such as growth rates or weather conditions, root mean square error (RMSE) is typically used to reflect the mean magnitude of prediction errors, with higher deviations given greater weights. Aside from metric choice, strict model assessment requires solid validation methods. Cross-validation, especially k-fold cross-validation, is a standard method where the data are divided into k subsets, and the model is trained and validated iteratively over various folds. This method reduces the risk of overfitting and provides a more accurate estimate of model performance across diverse data distributions, a critical factor in aquaculture where environmental and biological conditions vary over time and across different geographical regions (Owusu-Adjei et al., 2023). Model performance is typically optimized through a hyperparameter search. The hyperparameter learning rate, tree depth, or number of neurons in a neural network are not learned during training and must be predefined before training. Systematic search methods, such as Grid Search, which systematically tries all combinations within a specified set, and Random Search, which samples configurations at random, are used to determine the optimal values for hyperparameters. Advanced techniques, such as Bayesian optimization or genetic algorithms, are gaining popularity to augment tuning efficiency, particularly for deep models with wide hyperparameter spaces. Lastly, careful evaluation and optimization are also crucial for AI/ML model deployment to ensure accuracy, reliability, and robustness, thereby enabling more intelligent and environmentally friendly aquaculture practices (Passos & Mishra, 2022).

References

Abdullah, A. F., Man, H. C., Mohammed, A., Karim, M. M. A., Yunusa, S. U., & Jais, N. A. B. M. (2024). Charting the aquaculture internet of things impact: key applications, challenges, and future trend. *Aquaculture Reports, 39*. https://doi.org/10.1016/j.aqrep.2024.102358

Ahmed Murtaza, A., Saher, A., Hamza Zafar, M., Kumayl Raza Moosavi, S., Faisal Aftab, M., & Sanfilippo, F. (2024). Paradigm shift for predictive maintenance and condition monitoring from industry 4.0 to industry 5.0: A systematic review, challenges and case study. *Results in Engineering, 24*. https://doi.org/10.1016/j.rineng.2024.102935

Alkam, T., Van Benschoten, A. H., & Tarshizi, E. (2025). Reinforcement learning in artificial intelligence and neurobiology. *Neuroscience Informatics, 5*(3). https://doi.org/10.1016/j.neuri.2025.100220

Ashraf Rather, M., Ahmad, I., Shah, A., Ahmad Hajam, Y., Amin, A., Khursheed, S., … Rasool, S. (2024). Exploring opportunities of artificial intelligence in aquaculture to meet increasing food demand. *Food Chemistry X, 22,* Article 101309. https://doi.org/10.1016/j.fochx.2024.101309

Biazi, V., & Marques, C. (2023). Industry 4.0-based smart systems in aquaculture: a comprehensive review. *Aquacultural Engineering, 103.* https://doi.org/10.1016/j.aquaeng.2023.102360

Boyd, C. E., McNevin, A. A., & Davis, R. P. (2022). The contribution of fisheries and aquaculture to the global protein supply. *Food Security, 14*(3), 805–827. https://doi.org/10.1007/s12571-021-01246-9

Chandran, P. J. I., Khalil, H. A., Hashir, P. K., & Veerasingam, S. (2025). Smart technologies in aquaculture: an integrated IoT, AI, and blockchain framework for sustainable growth. *Aquacultural Engineering, 111.* https://doi.org/10.1016/j.aquaeng.2025.102584

Chilton, D., Hamilton, D. P., Nagelkerken, I., Cook, P., Hipsey, M. R., Reid, R., ... Brookes, J. (2021). Environmental flow requirements of estuaries: providing resilience to current and future climate and direct anthropogenic changes. *Frontiers in Environmental Science, 9.* https://doi.org/10.3389/fenvs.2021.764218

Dauda, A. B., Ajadi, A., Tola-Fabunmi, A. S., & Akinwole, A. O. (2019). Waste production in aquaculture: sources, components and managements in different culture systems. *Aquaculture and Fisheries, 4*(3), 81–88. https://doi.org/10.1016/j.aaf.2018.10.002

Essamlali, I., Nhaila, H., & El Khaili, M. (2024). Advances in machine learning and IoT for water quality monitoring: a comprehensive review. *Heliyon, 10*(6), Article e27920. https://doi.org/10.1016/j.heliyon.2024.e27920

Fini, C., Amato, S. G., Scutaru, D., Biancardi, S., Antonucci, F., Violino, S., ... Costa, C. (2025). Application of generative artificial intelligence in the aquacultural sector. *Aquacultural Engineering, 111.* https://doi.org/10.1016/j.aquaeng.2025.102568

Irshath, A. A., Rajan, A. P., Vimal, S., Prabhakaran, V. S., & Ganesan, R. (2023). Bacterial pathogenesis in various fish diseases: recent advances and specific challenges in vaccine development. *Vaccines (Basel), 11*(2). https://doi.org/10.3390/vaccines11020470

Liu, H., Chen, J., Dy, J., & Fu, Y. (2023). Transforming complex problems into K-Means solutions. *IEEE Transactions on Pattern Analysis and Machine Intelligence, 45*(7), 9149–9168. https://doi.org/10.1109/TPAMI.2023.3237667

Martinez-Porchas, M., & Martinez-Cordova, L. R. (2012). World aquaculture: environmental impacts and troubleshooting alternatives. *The Scientific World Journal, 2012,* Article 389623. https://doi.org/10.1100/2012/389623

Mohammed, E. A. H., Kovacs, B., Kuunya, R., Mustafa, E. O. A., Abbo, A. S. H., & Pal, K. (2025). Antibiotic resistance in aquaculture: challenges, trends analysis, and alternative approaches. *Antibiotics (Basel), 14*(6). https://doi.org/10.3390/antibiotics14060598

Niazi, S. K. (2023). The coming of age of AI/ML in drug discovery, development, clinical testing, and manufacturing: the FDA perspectives. *Drug Design, Development and Therapy, 17,* 2691–2725. https://doi.org/10.2147/DDDT.S424991

Owusu-Adjei, M., Ben Hayfron-Acquah, J., Frimpong, T., & Abdul-Salaam, G. (2023). Imbalanced class distribution and performance evaluation metrics: a systematic review of prediction accuracy for determining model performance in healthcare systems. *PLOS Digital Health, 2*(11), Article e0000290. https://doi.org/10.1371/journal.pdig.0000290

Passos, D., & Mishra, P. (2022). A tutorial on automatic hyperparameter tuning of deep spectral modelling for regression and classification tasks.

Chemometrics and Intelligent Laboratory Systems, 223. https://doi.org/10.1016/j.chemolab.2022.104520

Sáez-Arteaga, A., Viegas, I., Palma, M., Dantagnan, P., Valdebenito, I., Figueroa Villalobos, E., … Heyser, C. (2024). Impact of increasing temperatures on neuroendocrine and molecular responses of skeletal muscle and liver in fish: a comprehensive review. *Aquaculture Reports, 39.* https://doi.org/10.1016/j.aqrep.2024.102448

Saleh, A., Sheaves, M., Jerry, D., & Rahimi Azghadi, M. (2024). Applications of deep learning in fish habitat monitoring: a tutorial and survey. *Expert Systems with Applications, 238.* https://doi.org/10.1016/j.eswa.2023.121841

Sarker, I. H. (2021a). Deep learning: A comprehensive overview on techniques, taxonomy, applications and research directions. *SN Computer Science, 2*(6), 420. https://doi.org/10.1007/s42979-021-00815-1

Sarker, I. H. (2021b). Machine learning: algorithms, real-world applications and research directions. *SN Computer Science, 2*(3), 160. https://doi.org/10.1007/s42979-021-00592-x

Serrano, W. (2022). Deep reinforcement learning with the random neural network. *Engineering Applications of Artificial Intelligence, 110.* https://doi.org/10.1016/j.engappai.2022.104751

Shin, H. C., Roth, H. R., Gao, M., Lu, L., Xu, Z., Nogues, I., … Summers, R. M. (2016). Deep convolutional neural networks for computer-aided detection: CNN architectures, dataset characteristics and transfer learning. *IEEE Transactions on Medical Imaging, 35*(5), 1285–1298. https://doi.org/10.1109/TMI.2016.2528162

Shreesha, S., Pai, M. M. M., Verma, U., & Pai, R. M. (2023). Fish tracking and continual behavioral pattern clustering using novel Sillago Sihama vid (SSVid). *IEEE Access, 11,* 29400–29416. https://doi.org/10.1109/access.2023.3247143

Soussi, A., Zero, E., Sacile, R., Trinchero, D., & Fossa, M. (2024). Smart sensors and smart data for precision agriculture: a review. *Sensors, 24*(8). https://doi.org/10.3390/s24082647

Sriputhorn, K., Pitakaso, R., Matitopanum, S., Luesak, P., Khonjun, S., Kraiklang, R., … Gonwirat, S. (2025). Revolutionizing climbing perch disease management: AI-Driven solutions for sustainable aquaculture. *Smart Agricultural Technology, 10.* https://doi.org/10.1016/j.atech.2024.100746

Wägele, J. W., Bodesheim, P., Bourlat, S. J., Denzler, J., Diepenbroek, M., Fonseca, V., … Wildermann, S. (2022). Towards a multisensor station for automated biodiversity monitoring. *Basic and Applied Ecology, 59,* 105–138. https://doi.org/10.1016/j.baae.2022.01.003

Xu, Y., Liu, X., Cao, X., Huang, C., Liu, E., Qian, S., … Zhang, J. (2021). Artificial intelligence: A powerful paradigm for scientific research. *The Innovation, 2* (4). https://doi.org/10.1016/j.xinn.2021.100179

Yang, H., Feng, Q., Xia, S., Wu, Z., & Zhang, Y. (2025). AI-driven aquaculture: A review of technological innovations and their sustainable impacts. *Artificial Intelligence in Agriculture, 15*(3), 508–525. https://doi.org/10.1016/j.aiia.2025.01.012

Zarbakhsh, S., Fakhrzad, F., Rajkovic, D., Niedbała, G., & Piekutowska, M. (2025). Approaches and challenges in machine learning for monitoring agricultural products and predicting plant physiological responses to biotic and abiotic stresses. *Current Plant Biology, 43.* https://doi.org/10.1016/j.cpb.2025.100535

Zatsu, V., Shine, A. E., Tharakan, J. M., Peter, D., Ranganathan, T. V., Alotaibi, S. S., … Nayik, G. A. (2024). Revolutionizing the food industry: The transformative power of artificial intelligence-a review. *Food Chemistry X, 24*. https://doi.org/10.1016/j.fochx.2024.101867

Zhang, J., Li, C., Rahaman, M. M., Yao, Y., Ma, P., Zhang, J., ... Grzegorzek, M. (2022). A comprehensive review of image analysis methods for microorganism counting: from classical image processing to deep learning approaches. *Artificial Intelligence Review, 55*(4), 2875–2944. https://doi.org/10.1007/s10462-021-10082-4

Zhang, Y. J., Luo, Z., Sun, Y., Liu, J., & Chen, Z. (2023). From beasts to bytes: Revolutionizing zoological research with artificial intelligence. *Zoological Research, 44*(6), 1115–1131. https://doi.org/10.24272/j.issn.2095-8137.2023.263

6

Bioimaging and AI integration for aquaculture disease control and monitoring

Introduction

The farming of fish, crustaceans, mollusks, and aquatic plants is known as aquaculture, which significantly contributes to the global food production. With the growing global population, the demand for sustainable and efficient food production methods also increases. Since aquaculture plays an important role in meeting the demand for seafood, the aquaculture industry valued at over $200 billion annually, and the production will expand in the near future (FAO, 2020). Even though aquaculture is a potential industry, it faces many challenges which suppress its sustainability and efficiency. These challenges include effective resource management, disease prevention, environmental sustainability, and product quality. Addressing these challenges will ensure the long-term viability of the industry. Currently, the application of artificial intelligence (AI) provides promising solutions to overcome these challenges. Generally, AI, a branch of computer science that deals with the development of latent systems that accomplish human intelligence, necessitating responsibilities such as problem-solving, decision-making, and learning is primed for alteration of manufacturing. In this context, AI incorporation in aquaculture is anticipated to improve manufacture progressions, augment nursing and regulatory systems, and offer groundbreaking solutions to outdated farming glitches. Environmental monitoring and organization is one of the critical zones in which AI is integrated. The health and growth of cultivated species are directly wedged by water quality parameters such as temperature, pH level, oxygen content, and nutrient concentration. Traditional methods for nursing water quality involve more labour-intensive processes, regular manual selection, and periodic challenges. While AI-powered systems offer real-time, unceasing observation with the help of

Aquaculture with Artificial Intelligence. https://doi.org/10.1016/B978-0-443-33704-8.00009-4

devices, machine learning (ML) algorithms, and data analytics. These organizations detect fluctuations in water quality and also envisage the forthcoming trends that aid farmers to contrive effective actions to protect the cultivated species from harmful circumstances. For example, AI helps in predicting the algal blooms, which perhaps leads to oxygen reduction and a toxic situation, allowing farmers to take protective measures (Tarek et al., 2019). AI also benefits in adapting disease management in aquaculture. One of the hazardous threats to aquaculture is disease outbreaks, which cause economic losses and disturbance in global seafood stock chains. Avoiding extensive disease occurrences needs early detection of pathogens and diseases. Previous farmers trailed time-consuming and unsuccessful outmoded approaches of disease uncovering, such as pictorial examination and labor-intensive sampling in large-scale operations.

The premature signs of diseases in farmed species are perceived using AI skills, precisely ML and computer vision. AI replicas examine the images and videos of animals to sense slight fluctuations in the behavior, presence, and course of animals occurring as a result of the beginning of infection (Hernandez et al., 2020). Furthermore, historical health data is analyzed by AI, and it provides early warning on disease outbreaks to the farmers, which helps to take preventive measures such as adjusting water conditions or administering targeted treatments. AI also plays an essential part in enlightening the competence of alimentation practices in fish farming. Feeding is the most cost-effective process in fish farming. Therefore overfeeding and undernourishment lead to leftover, unfortunate development and enlarged environmental effect. Though outdated feeding approaches rely on fixed agendas, assessments built on the size and species of cultivated fishes, AI, on the other hand, permits exact feeding by analyzing real-time data of animal's development rate and environmental circumstances. To diminish the surplus and encourage better progress rates, the ideal amount and effectiveness of feed mandatory for each discrete cluster of animals is foreseen by ML algorithms (Baskaran et al., 2021). In addition, AI enhances feed formulation and certifies that the nutritive requirements of the farmed fishes are met and cuts the environmental footprint of feed manufacture. The amalgamation of AI into aquaculture improves the sustainability of the industry. Aquaculture operations have disadvantages such as environmental impact, particularly regarding water pollution, the use of antibiotics, and the depletion of wild fish stocks used as feed. Hence, implementation of AI-driven models helps farmers to condense water usage, minimize waste, and improve the effectiveness of energy use in aquaculture systems (Chen et al., 2022). Furthermore, the biodiversity of aqua farms is monitored by AI to ensure that local ecosystems are not destructively obstructed and there is no biodiversity loss. Even though AI has many potential applications, applying these technologies has its own tasks. First, the charge of AI systems, second, the requirement for high-quality data, and finally, the deficiency of technical experts in the industry. These contests should be considered to approve AI in aquaculture. Moreover, moral issues such as confidentiality, mechanization of labor, over dependence on AI must be well-thought-out. Also, these tasks AI integration in aquaculture makes an excessive influence which aids in meeting the request for sustainable, and well-organized seafood production. In this chapter, we will discover the numerous claims of AI in aquaculture, from environmental nursing to disease supervision and reserve optimization. We will also discuss the tasks and chances allied with the approval of AI in the industry and offer case studies that demonstrate the applied claims of AI technologies. As we move toward progressive and sustainable impending, the role of AI in aquaculture will be a transformative force, contributing to some of the critical challenges of the aquaculture industry.

Bioimaging technologies in aquaculture

In modern aquaculture, dynamic tools like bioimaging technologies allow noninvasive, real-time imaging and monitoring of the health, physiology, and environmental communications of aquatic organisms. These skills include magnetic resonance imaging (MRI), computed tomography (CT), ultrasound imaging, infrared thermography, fluorescence imaging, and hyperspectral imaging (Fig. 6.1). Each technique delivers exclusive data on biological processes, disease diagnosis, growth monitoring, feed idealization, and water quality assessment. For instance, Ultrasound imaging is used to screen the gonadal development and evaluate the reproductive position of broodstock without instigating stress or injury (Martínez-Pastor et al., 2011). Likewise, infrared thermography notices the pointers of contamination and wounds in fish and shellfish, such as localized inflammation and temperature oscillations (Sundberg et al., 2009). Besides the cost, MRO and CT scanning provide high-resolution internal images of the anatomy of the fish, which is essential in research and broodstock management (Grosenick et al., 2007). The amalgamation of bioimaging data, AI, and ML algorithms enriches the analytical capabilities of aquaculture systems. AI can also be

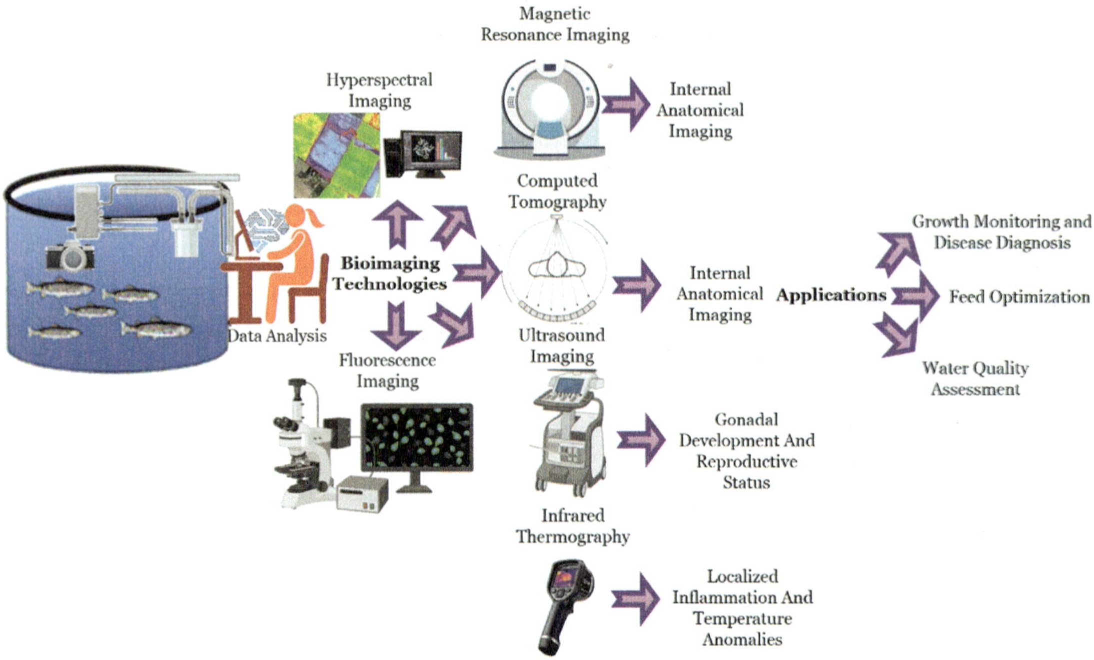

Figure 6.1 Various bioimaging technologies and their applications in aquaculture.

trained to detect the early symptoms of diseases, behavioral abnormalities, highly accurate classification of species and their developmental stages. This, in turn, helps in automating health diagnostics and minimizing human error (Rani et al., 2022). Hyperspectral imaging, in combination with AI, rapidly assesses the freshness of fish and their quality by analyzing spectral signals linked with biochemical changes (Pu et al., 2015). The role of bioimaging technologies enhanced by AI is prominent as the aquaculture industry is focusing on precision farming. These innovations not only support sustainability but also reduce the reliance on invasive techniques. Also, these techniques improve welfare standards and enable better management decisions. It also holds the potential to transform the future aquaculture, particularly in optimizing production efficiency, ensuring animal welfare, and maintaining ecological balance.

Role of artificial intelligence in bioimaging

The domain of bioimaging has been reformed by AI, which offers potential applications in aquaculture by improving the precision, efficiency, and automation of image-based analysis. The bioimaging AI, along with ML and deep learning (DL) algorithms, ensures accurate interpretation of biological images for assessing the health, morphology, and development of farmed organisms. Outdated image examination is frequently labor-intensive, in which physical clarification leads to unpredictability and inadequate scalability. AI systems customise bulky datasets, and notice complicated outlines in imaging data to enable real-time intensive care and decision making (Esteva et al., 2019). In aquaculture tasks such as detection of fish species, parasite detection, wound arrangement, and progress dimension are successfully carried out with AI assistance (Park et al., 2020). Convolutional neural networks (CNNs) are a class of DL replicas that validates high precise subdivision and sorting features in histologic and microscopic images of aquatic organisms (Li et al., 2021).

Moreover, AI models also process multimodal imaging data, such as fluorescence, MRI, or hyperspectral images, to deliver profounder perceptions into physiological and pathological settings in place of aggressive techniques (Gulshan et al., 2016). To boost the unceasing nursing of fish stock health, to permit early disease investigation and to lessen mortality rates, the Internet of Things (IoT) is combined with AI, and the imaging systems in aquaculture farms are computerized (Nayak et al., 2022). In addition, AI-driven bioimaging can monitor destructive algal blooms and impurities over satellite and drone-based imaging (Kamilaris & Prenafeta-Boldú, 2018). To sum up, the role of AI in bioimaging suggestively pays to sustainable aquaculture by reducing human blunders, utilizing the reserve properly, and fostering effective health management.

Integration of bioimaging and AI: System architecture

The combination of bioimaging and AI provides an effective transformation in advancing precision aquaculture by developing intelligent system architectures. Bioimaging technologies such as digital imaging, hyperspectral imaging, X-ray, Ultrasound, and fluorescence microscopy help in discovering noninvasive, real-time visual data which is significant for observing fish health, behavior, growth designs, and environmental circumstances. When AI procedures are joined with ML, DL, and computer vision, imaging techniques empower the automatic investigation and explanation of wide and complex data streams, instantaneously enhancing the decision-making in aquaculture organizations (Kumar et al., 2021). In general, a system architecture is an integration of several key modules, including data acquisition through bioimaging devices, preprocessing to reduce noise and image improvement, feature extraction using CNNs, and the classifying or predicting models to detect diseases, assess biomass, track movement patterns (Fig. 6.2). Furthermore, the integration with IoT devices and cloud-based platforms enables data transfer, remote monitoring, and operational scalability (Rahman et al., 2022). For example, AI models trained on annotated bioimaging datasets identify early signs of parasitic infections or skeletal deformities in fish larvae, which are often challenging for human observers to detect visually (Sharma et al., 2020). Such systems are embedded in smart aquaculture infrastructures, which enable continuous and autonomous surveillance, which in turn minimizes labor costs and increases productivity. The incorporation of feedback loops and adaptive learning capabilities in these systems evolved the response to variability in species, changes in the environment, and farming practices. This integrated architecture not only contributes to sustainable aquaculture but also aligns with the global objectives of smart farming and digital reformation in the fisheries sector. Interdisciplinary collaboration among aquaculturists, data

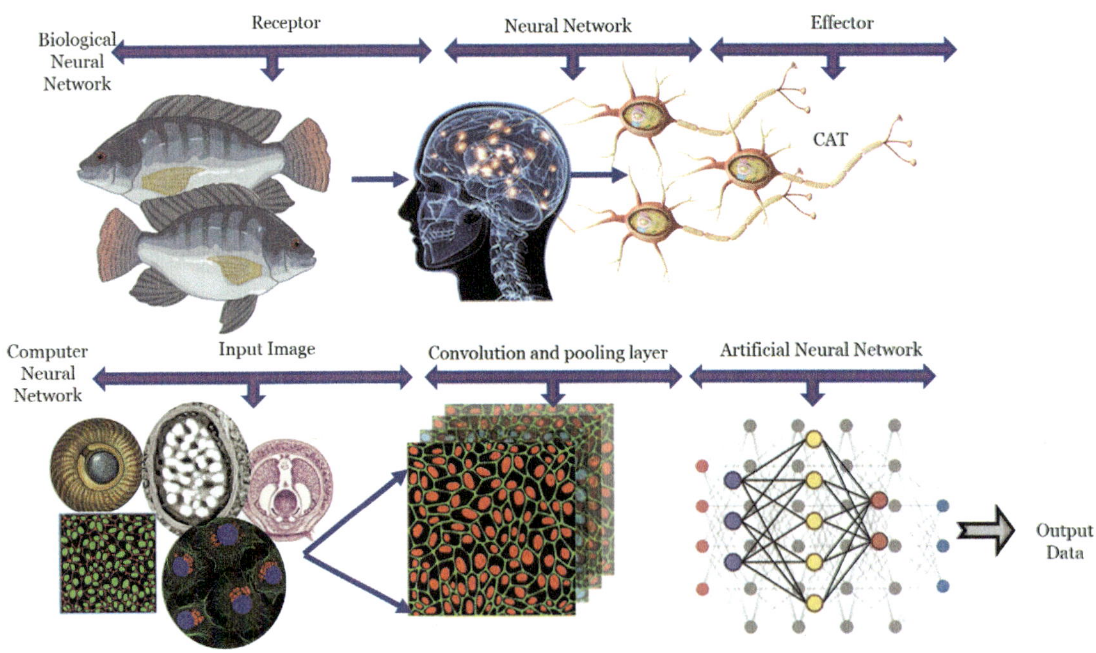

Figure 6.2 Convolutional neural networks with neuro imaging techniques in aquaculture.

scientists, and engineers is essential to overcome the challenges of this progressing field. Those challenges include data standardization, model generalization, and deployment in real-world aquatic environments (Li et al., 2023). Thus the aquaculture practice is redefined by the synergy of bioimaging and AI within a strong system architecture ensuring efficiency, sustainability, and food security.

Disease detection and monitoring is one of the reformative application of AI in aquaculture, where early diagnosis and real-time surveillance are necessary to maintain aquatic animal health and to reduce economic losses. Microscopic examination, biochemical assays, and histopathology fall under traditional methods of disease diagnosis, which are not only reliable but also time-consuming, labor-intensive, and require skilled personnel (Ramos et al., 2021). AI provides a strong alternative by involving image recognition, ML, and data analytics to rapidly identify disease symptoms and monitor health status. Computer vision techniques supported by CNNs analyzing fish behavior, skin lesions, gill condition, and color variations from underwater footage to detect early signs of bacterial gill disease or parasitic infections is an example (Qin et al., 2020) (Fig. 6.3). ML algorithms trained on historical health data, environmental parameters, and feeding behavior predict potential disease outbreaks with high

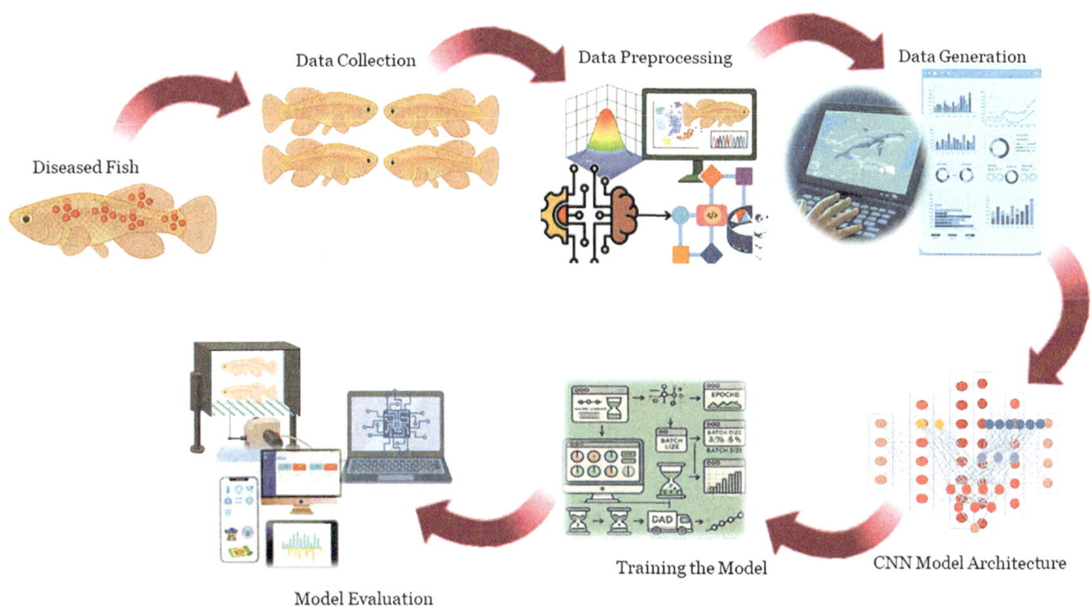

Figure 6.3 Flowchart of fish disease detection using CNN in aquaculture.

accuracy, enabling farmers to take proactive measures (Li et al., 2022). AI-based biosensors and IoT stages further improve real-time nursing by uninterruptedly assembling and conveying water quality data, including key temperature, pH, ammonia, and dissolved oxygen levels, which are key pointers of stress and vulnerability in aquaculture species (Alam et al., 2021). Mixing AI with these technologies ensures precision aquaculture, dropping dependence on antibiotics and chemicals, therefore endorsing justifiable practices.

Besides, the incorporation of natural language processing (NLP) and expert systems helps in mechanical reporting and supports decision-making, which is valuable in distant aquaculture actions (Xie et al., 2023). As well as present tasks with the necessity for huge, marked up datasets and the intricacy of biological inconsistency, the rapid progression in AI models and cloud computation is poised to revolutionize disease supervision in aquaculture. These smart systems not only rise proficiency and yield but also subsidize meaningfully to animal welfare and food security in the face of worldwide aquaculture expansion.

Implementation in farm settings

The conversion from the usage of old-style methods to powerful, meticulous agrobusiness and sustainable development is the outcome of the employment of AI in aquaculture. The participation of AI in daily processes allows real-time intensive care, decision-making, and extrapolative analytics, and recovers effectiveness and reduces operative dangers. AI-powered sensors, computer vision systems, and IoT strategies are used to display serious parameters such as water temperature, dissolved oxygen, pH, ammonia levels, and fish behavior at the farm level. The primary finding of irregularities and disease occurrences is probable by these tools, thus it lessens deaths and financial sufferings. ML algorithms analyze past and real-time data to augment feeding agendas, diminish feed waste, and improve progress rates (Li et al., 2021). In recirculating aquaculture systems (RAS), AI is utilized to control ventilation, percolation, and water exchange procedures to confirm optimal environmental situations for the cultivated aquatic organisms. Drones and self-governing underwater automobiles are reinforced by AI-based tomography, ease stock assessment and estimation of health, which delivers high tenacity understandings without disrupting the aquatic environment. In addition, decision support systems

(DSS) with AI models help farm administrators to make informed choices on harvesting, stocking densities, and biosecurity protocols. Despite their benefits, AI applications in agricultural farms need an acceptable substructure, appropriate internet connectivity, skilled technical workers, and well-organized data management. AI designers, aquaculture specialists, and farmers collaborate with each other to advance operator-friendly tools customized to the precise needs of discrete farms. Further, monetary inducements and policy support surge the approval of AI technologies in aquaculture. Ultimately, the grouping of AI and aquaculture at the agricultural level opens the way for bright, strong, and eco-friendly possible farming systems to meet the increasing global demand for aquatic food (FAO, 2020).

Challenges and limitations

Alterations in aquaculture, such as disease recognition, food optimization, water quality management, and total farm organization, have been made by the implementation of AI into aquaculture. However, several challenges and limitations deter its implementation and effectiveness. Data availability and quality are the primary challenges. AI systems rely heavily on large volumes of high-quality specific data to perform accurately. In aquaculture, such datasets for species-specific behaviors, disease markers, and environmental parameters remain difficult due to the uneven nature of data collection and lack of standard protocols across farms (Gupta et al., 2021). In addition, various range of aquatic species and farming environments makes it difficult to develop universally applicable AI models. A model working for one species in a particular region may not work appropriately in another location, which requires localized training and customization of AI tools (Kim et al., 2022).

In developing countries, there are significant barriers, such as limited cost and infrastructure, reliable internet connectivity, power supply, and modern hardware such as sensors, underwater cameras. Lack of technical expertise among aquaculture practitioners is another concern. Most of the farm operators have less knowledge on AI or data science, which results in a gap between technological advancements and their handling in farms. This, in turn, resists the adoption of AI-driven tools because of complexity and job displacement fear. Furthermore, moral and governing rules concerning data privacy and ownership, as well as wide-ranging agendas to administer AI use in aquaculture, must be tracked (Rao, 2026). Also, AI systems fight with changeable natural events such as climate alteration and pathogen epidemics, which restrict the dependability of prognostic models. Such encounters signify the necessity for interdisciplinary partnership, strong data distribution mechanisms, and user-centric AI tools to utilize the capability of AI in aquaculture.

Future prospects and innovations

The contribution of AI to aquaculture has the potential to revolutionize traditional fish farming practices and provide intelligent and data-driven solutions. AI in aquaculture provides a sustainable pathway to enhance productivity, ensure animal welfare, and reduce environmental impacts. In short, AI technologies will dominate the aquaculture value chain from hatchery to harvest. Precision aquaculture, along with ML algorithms, computer vision, and IoT sensors, offers real-time monitoring of the farm, which includes water quality parameters, fish behavior, and health status. Accurate assessments of stock density, disease detection, and timely interventions will be enabled by innovations in imaging and drone-based surveillance. Predictive analysis is important to optimize feed utilization, observe growth rates, and determine harvest times, which in turn leads to cost savings and increased yield.

In addition, blockchain technology enhances traceability and transparency when combined with AI, and it builds consumer trust and ensures compliance with regulatory standards. To help farmers adapt to the changes in environmental conditions, AI aids in ecosystem modeling and climate risk assessment on a large scale. The establishments of autonomous feeding systems, robotic net cleaners, and biosecurity systems driven by AI will redefine operational efficiency in aquaculture farms. Future research focuses on the interpretability and strong AI models to remain effective under various dynamic aquaculture environments. Association among researchers, technologists, and industry stakeholders will accelerate innovation and provide responsible usage of AI. Finally, the joining of AI in aquaculture offers an intelligent, resilient, and environmentally responsible future for the blue economy.

Conclusion

The AI invasion in aquaculture marks a turning point in the way aquatic farming is conceptualized, managed, and sustained. The global demand for seafood increases, and so traditional aquaculture practices undergo significant challenges such as disease outbreaks, environmental degradation, inefficient resource utilization, and inconsistent production yields. AI, with its potential in data analysis, pattern recognition, and predictive modeling, provides effective solutions to the above issues. To conclude, this chapter underlines the position of AI-processed tools in ML procedures, computer vision systems, feed idealization and organization, water quality regulation, disease diagnosis, and inclusive farm administration. These progressions endorse sustainable methods such as waste reduction, negligible ecological footprint in aquaculture processes. Additionally, the grouping of AI allows real-time decision-making, which advances the confrontation of aquaculture systems in contradiction to varying environmental settings. Precisely, the fruitful employment of AI in aquaculture needs a multidisciplinary method with association among marine biologists, data scientists, engineers, and policy makers. The operative adoption of these inventiveness requires capacity building, infrastructure expansion, and ethical contemplations for data privacy and available technology. To sum up, as this sector progresses, AI will become an exceptional tool in aquaculture, which brings it into line with global goals for food security, environmental conservation, and financial development. Thus the future of this aquaculture sector depends AI-driven inventions that help in origination, competence, and sustainable fish farming methods across numerous aquatic ecosystems.

References

Alam, M. A., Rahman, M. A., & Lee, S. H. (2021). Smart aquaculture system with monitoring and prediction of fish behavior. *Aquaculture Engineering, 95*, Article 102169.

Baskaran, R., Muthukumar, T., & Rajendran, R. (2021). Precision feeding in aquaculture using artificial intelligence: A review. *Aquaculture Research, 52* (7), 2879–2895.

Chen, Y., Chen, Y., & Zhang, W. (2022). Optimizing resource use in aquaculture: A machine learning approach. *Aquaculture Economics and Management, 26* (4), 463–485.

Esteva, A., Robicquet, A., Ramsundar, B., et al. (2019). A guide to deep learning in healthcare. *Nature Medicine, 25*(1), 24–29.

FAO. (2020). The state of World fisheries and aquaculture 2020: Sustainability in action. *Food and Agriculture Organization of the United Nations*, 1–244.

Grosenick, L., Smith, B. H., & Anderson, D. J. (2007). Imaging neural activity in unrestrained animals. *Current Opinion in Neurobiology, 17*(5), 593–600.

Gulshan, V., Peng, L., Coram, M., et al. (2016). Development and validation of a deep learning algorithm for detection of diabetic retinopathy in retinal fundus photographs. *JAMA, 316*(22), 2402–2410.

Gupta, A., Gupta, S. K., Priyam, M., Siddik, M. A., Kumar, N., Mishra, P. K., Gupta, K. K., Sarkar, B., Sharma, T. R., & Pattanayak, A. (2021). Immunomodulation by dietary supplements: A preventive health strategy for sustainable aquaculture of tropical freshwater fish, *Labeo rohita* (Hamilton, 1822). *Reviews in Aquaculture, 13*(4), 2364–2394.
Hernandez, A., Tarek, M., & Ahmed, M. (2020). AI-based disease detection in aquaculture. *Aquatic Health Journal, 23*(1), 42–54.
Kamilaris, A., & Prenafeta-Boldú, F. X. (2018). Deep learning in agriculture: A survey. *Computers and Electronics in Agriculture, 147,* 70–90.
Kim, J., Kim, Y., Park, S. E., Kim, T. H., Kim, B. G., Kang, D. J., & Rho, T. (2022). Impact of aquaculture on distribution of dissolved organic matter in coastal Jeju Island, Korea, based on absorption and fluorescence spectroscopy. *Environmental Science and Pollution Research, 29*(1), 553–563.
Kumar, P., Gupta, D., & Banerjee, S. (2021). Applications of artificial intelligence in aquaculture: A review. *Aquaculture Engineering, 94,* Article 102396.
Li, Y., Huang, Y., Zhang, S., et al. (2021). Automated microscopic image analysis for fish histology using deep learning. *Aquaculture Reports, 19,* Article 100582.
Li, Y., Zhang, X., & Li, D. (2022). Machine learning techniques for fish disease diagnosis: A review. *Computers and Electronics in Agriculture, 195,* Article 106812.
Li, X., Zhao, Y., & Chen, M. (2023). Artificial intelligence for sustainable aquaculture: Emerging trends and future perspectives. *Journal of Marine Science and Engineering, 11*(1), 42.
Martínez-Pastor, F., Guerra, C., Álvarez, M., & Anel, L. (2011). Application of imaging technologies in fish reproduction and development. *Theriogenology, 76*(9), 1465–1478.
Nayak, S. K., Sundaray, J. K., & Swain, P. (2022). AI and IoT integration in aquaculture: A paradigm shift. *Aquaculture International, 30*(3), 1271–1290.
Park, S. H., Kim, D., Lee, D. H., & Jeong, W. (2020). Deep learning-based fish species classification using underwater imagery. *Sensors, 20*(20), 5969.
Pu, H., Sun, D.-W., Ma, J., & Cheng, J.-H. (2015). Classification of fresh and frozen-thawed fish fillets using near-infrared hyperspectral imaging with multivariate analysis. *Food and Bioprocess Technology, 8,* 1656–1667.
Qin, J., Li, H., & Li, Y. (2020). Computer vision-based fish disease detection: A review. *Aquaculture, 526,* Article 735357.
Rahman, M. M., Islam, M. S., & Ali, M. Y. (2022). IoT and AI-enabled smart aquaculture: Architecture and future directions. *Computers and Electronics in Agriculture, 196,* Article 106867.
Ramos, F., Silva, C., & Oliveira, R. (2021). Challenges in fish disease diagnostics and AI-driven solutions. *Frontiers in Veterinary Science, 8,* Article 657891.
Rani, S., Rathore, S. S., & Rajawat, A. S. (2022). Application of AI in aquaculture: A review. *Aquaculture International, 30,* 2503–2521.
Rao, P. K. (2026). Emerging AI and ML tools for rapid diagnosis of bacterial diseases in aquaculture. *PriMera Scientific Medicine and Public Health, 8,* 03–12.
Sharma, R., Thomas, J. V., & Bhosale, S. (2020). Deep learning-based bioimaging in aquaculture: Applications and challenges. *Aquaculture Reports, 17,* Article 100374.
Sundberg, P., Gustavsson, E., & Löfstedt, H. (2009). Infrared thermography in veterinary medicine. *Thermology International, 19*(3), 105–110.

Tarek, M., Shih, W., & Li, F. (2019). Early detection of harmful algal blooms in aquaculture using AI-based systems. *Journal of Aquaculture Engineering, 11* (3), 119–133.
Xie, S., Zhou, X., & Wang, Y. (2023). Applications of natural language processing in smart aquaculture systems. *Journal of Aquaculture Research & Development, 14*(2), Article 100081.

7

AI integrating drone technologies for feeding technology

Introduction

The ambiguity of artificial intelligence (AI) and drone technologies has transformed abundant industries, with agriculture and animal farming standing at the vanguard of this revolution. Among the developing applications, AI-combined drone technologies for feeding systems signify a pioneering innovation meant to improve productivity, decrease labor, and advance the meticulousness of food dispersal in both terrestrial and aquatic farming systems. The mixture of autonomous aerial systems with intelligent algorithms permits instantaneous policy-making, adaptive feeding plans, and site-precise resource management, redesigning contemporary feeding performs. In current years, the increasing worldwide demand for food, united with labor deficiencies in the agricultural sector, has accentuated the requirement for automation and groundbreaking solutions (FAO, 2022). Outmoded feeding procedures, whether for livestock, poultry, or aquaculture, are labor-intensive and susceptible to inadequacies including overfeeding, underfeeding, and food wastage, all of which affect the economic feasibility and environmental longevity of farming applies (Shamshiri et al., 2018). Drone technology, when ambitious by AI, provides a favorable outcome by allocating feed self-sufficiently based on synchronized examination of environmental circumstances, animal behavior, and alimentary requirements. AI algorithms, with machine learning (ML) and deep learning (DL), authorize drones to progress data from involved detectors and cameras, identify alimentation zones, and make conversant conclusions about feed quantity and distribution patterns (Zhang et al., 2021). As in aquaculture, AI-ambitious drones investigate the key parameters of water quality, distinguish fish behavior, and exactly distribute feed, hence abating left-over and enhancing growth

Aquaculture with Artificial Intelligence. https://doi.org/10.1016/B978-0-443-33704-8.00010-0

(Wang et al., 2020). In livestock management, drones equipped with computer vision systems screen forms' movement and feeding behavior, ensuring the well-adjusted delivery of nutrients to precise groups or individuals based on the well-being system of dimension and alimentary requirements (Yang et al., 2022). Moreover, the merger of Geographic Information Systems (GIS) and remote perception technologies with AI-determined drones further expands spatial precision and meticulousness in feed delivery. These technologies permit the mapping of huge agricultural zones, documentation of vegetation indices, and examination of animal distribution, which are crucial for scrupulous feeding in prairie-based systems (Rejeb et al., 2022). Besides, cloud computing and Internet of Things (IoT) platforms equilibrium the drone performances by authorizing combined data collection, storage, and examine across coupled devices, providing ranchers with rapid perceptions and regulate over feeding measures (Kumar et al., 2019). The enlargement and implementation of AI-united drone systems in feeding technology also guarantees to maintainable of agriculture. These systems support in declining environmental footprints by optimizing feed usage, reducing methane production from livestock due to improved diet management, and curtailing nutrient excess into environments (Tilman et al., 2011). Besides, they promote animal well-being by sustaining reliable feeding programs and punctually identifying entities' needs that require consideration, which is predominantly esteemed in wide-ranging operations where labor-intensive monitoring is impracticable. Contempt the recurrent advantages, several encounters must be addressed for the wide-ranging adoption of this technology. Technical problems, as well as battery life, payload limitations, and severe environmental circumstances, disturb drone performance. As well, large early investment costs, lack of technical expertise among ranchers, and regulatory restraints pose fences to amalgamation, predominantly in emerging countries (Balafoutis et al., 2017). Nevertheless, ongoing research and progress efforts, joined with supportive policy frameworks, are predictable to alleviate these encounters and substitute wider adoption.

Principles of automated feeding systems

The principles of automated feeding systems are grounded in the merging of meticulous agriculture, robotics, and artificial intelligence (AI), empowering well-organized, appropriate, and adaptive distribution of nutrients in livestock and aquaculture

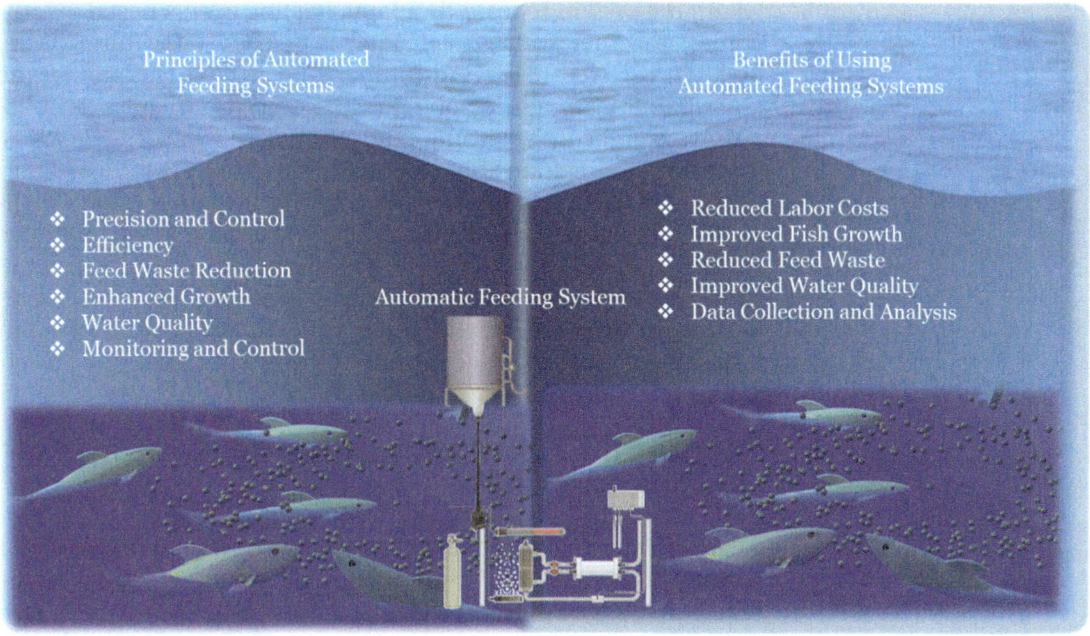

Figure 7.1 Major principles and benefits of automated feeding systems.

farming. These systems are designed to augment feeding procedures by plummeting labor, minimalizing feed leftovers, and improving animal well-being and output (Fig. 7.1). An essential principle comprises instantaneous sensing and data acquisition using progressive radars and vision systems to monitor animal behavior, body condition, and environmental parameters (Berckmans, 2017).

Data from these fundamentals is accomplished by means of AI algorithms, together with machine learning and deep learning models, which permit extrapolative analytics to adjust the level, type, and efficiency of feed (Liakos et al., 2018). Robotics is executed through programmable feeders, robotic arms, or drones, which are merged through Internet of Things (IoT) platforms to support whole communication and remote control (Wolfert et al., 2017). Drone-associated feeding systems, specifically, work with geolocation technologies and spatial mapping to distribute feed with precision in extensive farming atmospheres, together with aquaculture ponds or unconfined livestock fields. These drones were automated to trail predefined routes or, with dynamism, alter their routes based on immediate analytics, guaranteeing even feed distribution and preventing overfeeding or underfeeding (Zhou et al., 2020). Besides, feedback loops are crucial, where the system repeatedly adapts feeding events based on outcomes, together with growth rate, feed conversion ratio (FCR), and animal well-being indicators. Ethical contemplations and sustainability are also essential to the design principles, emphasizing insignificant environmental stimulus, a compressed carbon footprint, and responsible resource utilization. Ethical considerations and sustainability are also important to the design principles, emphasizing slight environmental influence, an abridged carbon footprint, and responsible resource utilization. The amalgamation of AI in automated feeding not only progresses operational effectiveness but also aligns with the wider goals of smart farming and longevity agriculture in the era of trade.

Role of AI in decision-making and optimization

Artificial Intelligence (AI) plays an essential role in decision-making and optimization progressions, predominantly in the amalgamation of drone technologies for feeding systems in agricultural and aquacultural atmospheres. AI algorithms authorize drones to operate autonomously, adapt to fluctuating environments, and make instant decisions based on data collected through numerous radars. These technologies are particularly beneficial in precise feeding, where enlarging the amount, timing, and location of feed delivery is crucial to taming output and reducing waste. Machine learning (ML) models analyze primary data, conservation parameters, and animal behavior patterns to normalize the optimum feeding schedule and conduct. For example, in aquaculture, AI-combined drones screen water quality, fish density, and feeding responses, then vigorously adjust feeding rates to assurance competent source use (Liakos et al., 2018). Policymaking outlines determined by AI also sustenance a means of augmentation for drones, reducing energy consumption and enhancing attention competence. Techniques including genetic algorithms, strengthening learning, and horde intelligence have visible significant potential in improving drone flight paths, load capacities, and developing numerous drone swarms for coordinated operations (Zhang et al., 2020). Furthermore, integrating AI into policymaking supports predictive analytics, enabling the system to forecast feeding requirements and identify anomalies such as equipment failures or unexpected environmental changes. This predictive ability not only boosts operational efficiency but also reduces human intervention, labor costs, and risks associated with labor-intensive investigations in remote or hazardous locations. As AI technologies continue to evolve, their application in drone-assisted feeding systems is expected to become more sophisticated, allowing real-time optimization that adapts to changing field conditions. This synergy between AI and drone technology represents a transformative approach in smart agriculture and aquaculture, contributing to longer-lasting performance and improved yields outcomes (Kamilaris & Prenafeta-Boldú, 2018).

Drone technology for monitoring and feed delivery

AI- empowered drone systems are revolutionizing both agriculture and aquacultural feed delivery, making a determinative development in longevity food production. These Unmanned Aerial Vehicles (UAVs) automatically supervise continuously in real-time feed dispersion, optimizing precision, operational efficiency, and environmental footprints. Drone with multispectral or thermal imaging technologies assesses the key water quality indicators, including temperature, dissolved oxygen, and algal bloom presence, enabling appropriate and adaptive feed modulation in aquaculture (Liu et al., 2020) (Fig. 7.2). Drones in

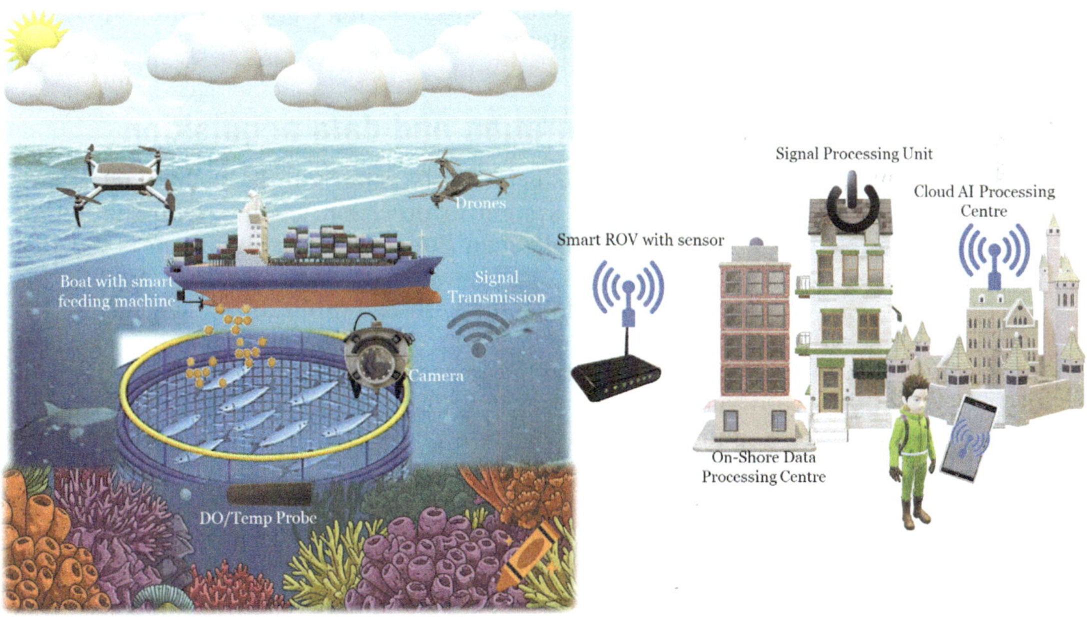

Figure 7.2 Advanced drone technology for monitoring and feed delivery in aquaculture systems.

agriculture used in mapping fields; to identify the nutrient deficient zones through this it delivers the fertilizer to nutrient depleted field and reduce the wastage and enhance the yield (Zhang & Kovacs, 2012). The AI-generated data increases the drone's decision-making capacity in real time through image recognition, environmental data, and historical trends. The machine learning algorithms revise the feeding schedule and amounts using predictive analysis. It operates by responding to animal behavior and environmental circumstances, which effectively decreases the overfeeding and underfeeding (Shahbazi et al., 2021). As well, drones autonomously disperse the feed in unfavorable environmental conditions, resulting in a reduction of manual interference. The radar with GPS and the Internet of Things enables uninterrupted communication between drones and control units, which establishes a closed loop in automated feeding infrastructure. This system improves animal welfare, environmental compliance, and operational efficiency. The ongoing advancements in drone technology, such as battery technology, lightweight materials, and AI algorithms, overcome the challenges, such as limited flight time, payload constraints, and regulatory restrictions. This merged artificial intelligence is a longevity route in the feed dispersal system since it is durable, intuitive, and environmentally sustainable networks (Gutiérrez et al., 2021; Tripicchio et al., 2015). This industrialized meticulousness feeds technologies, making agriculture and aquaculture a progressive system.

Sensor integration and data acquisition

Artificial intelligence-integrated drone technologies are a more advanced system in precision feeding, with the addition of sensors and data acquisition. This system enhances the ability of the drone in agricultural and aquacultural systems to make real-time decisions for resource utilization through environmental parameters. This integrated drone system has multispectral, hyperspectral, infrared, thermal, LiDAR, ultrasonic, and visual RGB cameras. Capture the animal behavior, feeding responses in the aquacultural system; on the other hand, it analyzes the crop health, soil moisture in the agricultural system. These multiple sensors are merged with unmanned aerial vehicles (UAVs) by sophisticated hardware, software interfaces to get seamless communication and synchronized data processing. The targeted feeding and mapping will be achieved with the integration of global navigation satellite systems (GNSS) and inertial measurement units (IMUs) to ensure precise geolocation and orientation data (Zhang & Kovacs, 2012). The onboard microcontrollers or edge computing models process the inputs in real-time, and they also manage the AI-driven feeding system data acquisition. This system assesses the acquired data and generates actionable insights using machine learning algorithms. For example, stress among the livestock and fish population is monitored using a thermal sensor; as a result, the feeding is scheduled based on their need by an AI-integrated system. Similarly, the nutrient-depleted area was captured by a multispectral imaging sensor, which helps to analyze the malnutrition in crops (Sankaran et al., 2015). Furthermore, the merging of wireless modules, including Wi-Fi, Bluetooth, or LoRaWAN, into this system transmits the data to the central processing unit or cloud-based platforms for further analysis, storage, and visualization. The victory of AI-integrated feeding drones is achieved through the quality and resolution of sensor data. Hence, the fusion of data techniques leads to combining data from several sensors to make an accurate and reliable real-time decision-making process. This fusion takes place based on the application and computational resource availability at different levels, such as the raw data level, feature level, or decision level. The AI algorithms, especially deep learning models, use the deciphered complex patterns in the fused data to enhance the predictive analysis in feeding schedules, quantity optimization, and environmental impact assessments (Yang et al., 2021).

AI-drone communication systems and frameworks

Artificial Intelligence (AI)-combined drone communication systems have developed as essential enablers in progressing feeding technologies, predominantly in the agricultural and aquacultural sectors (Fig. 7.3). These systems empower autonomous drones to interconnect efficiently with crushed control positions, radars, IoT networks, and other drones, guaranteeing instantaneous data broadcast and conversant policymaking. The introductory framework encompasses a multidimensional architecture, such as data acquisition, communication procedures, AI processing units, and actuation modules. At the fundamental of this system is the practice of AI algorithms, including convolutional neural networks (CNNs) and strengthening learning for object discovery, behavioral examination of livestock or aquatic animals, and forecast of feeding necessities based on

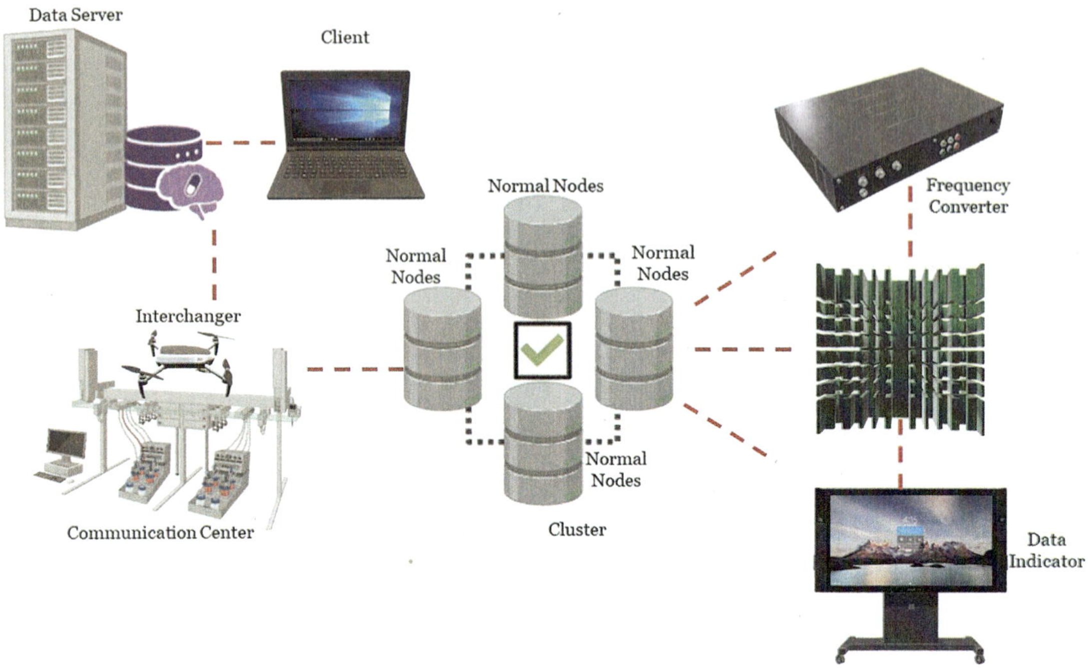

Figure 7.3 Artificial intelligence (AI)-integrated drone communication systems in aquaculture system.

environmental circumstances and animal behavior (Ghosh & Rihel, 2020). Communication frameworks characteristically influence wireless technologies, including Wi-Fi, 4G/5G, ZigBee, or LoRaWAN, which rely on the assortment and bandwidth necessities, guaranteeing seamless communication between drones and radar nodes (Sharma et al., 2025). These frameworks must account for potential, bandwidth optimization, fault tolerance, and cybersecurity. Edge computing plays a vital role by permitting drones to process wide volumes of visual and sensory data locally, minimalizing postponements and plummeting the need for continuous cloud connectivity (Khan et al., 2021). Besides, swarm intelligence frameworks have been combined, allowing coordinated drone actions throughout extensive feeding tasks, including uniformly distributing feed across aquaculture ponds or agricultural fields. In such systems, AI-ambitious policymaking confirms resource-competent operations, regulating feeding timetables in instantaneous based on animal demand and environmental parameters, including temperature, pH, or dissolved oxygen. Additionally, AI-integrated drone models change their feeding schedules using their adaptive learning process, which improves the feeding and reduces the wastage (Yue & Shen,

2022). In the future, the frameworks will be incorporated with blockchain to get secure data transmission digital twins' model to optimize the feeding before they are implemented. These integrated AI drone communication frameworks lead to the feeding system into sustainable, scalable, and productive technologies in precision agriculture and aquaculture.

Case studies and real-world applications

The integrated Artificial Intelligence (AI) and drone technology plays a major role in agricultural and aquacultural feeding systems, especially in automated systems. During the real-time operations, this system enhances food security, optimizes resource utilization, and reduces labor needs. The notable case study by Zhang et al. (2021) demonstrated that AI-integrated drones with multispectral imaging and deep learning algorithms assess the crop health and nutrient deficiencies in large area paddy fields in China. Similarly, Wageningen University in the Netherlands did a project called Pilot. In this project, they made automated AI-integrated drones to analyze the livestock movement and feed dispense in dairy farms, which reduces the feed waste and improves animal well-being (van der Voort et al., 2023). In India, Fasal and AgriBotix startups made AI-integrated drones to analyze the soil moisture and crop stress using machine learning algorithms. This system optimizes the feeding quantity and schedules. In the aquaculture field, the AI-integrated underwater drones are used to assess the feeding behaviors in fish, based on which they reschedule the feeding dispensation in real time, as done by Kim et al. (2023). In addition, John Deere and DJI multinational corporations collaborated to make AI-integrated drones with mapping, predictive analytics for yield outcomes, and the execution of precise farming with minimal labor. This system leads to smarter technology in the feeding system through analysis of mapping, environmental assessment, and optimization of feeding in both aquatic and agricultural sectors. This technology is more mature when introducing IoT devices, and blockchain enhances this system in farming practices.

Challenges and limitations in implementation

This automatically integrated artificial intelligence (AI) with drone technologies in the feeding system is an advanced one, but it faces several challenges and limitations in both the aquacultural and agricultural sectors during the implementation process. The main challenge of this technology is technical issues in AI algorithms, real-time analysis when it works in unstructured environments such as open fields and aquacultural ponds. This happened to the system due to the environmental changes in the open field, such as wind, temperature, climate changes, and unpredictable animal behaviors. To resolve this, it needs adaptive AI models with high-tech drone hardware (Zhang et al., 2020). The battery, payload capacity of the drone are another notable challenge in this system when it comes to the distribution feed in the large-scale sector (Shi et al., 2022). The data acquisition and processing are another notable challenge when detecting object recognition, feed distribution, and high-quality data set monitoring, which is unavailable or expensive to obtain. The cost and development, maintenance of this system are a barrier to rural farmers. Moreover, regulatory problems surrounding drone usage, predominantly regarding airspace permissions and privacy concerns, differ between countries and frequently delay execution in remote or highly controlled areas. From a moral standpoint, difficulties also arise concerning the dislodgment of manual labor and the socioeconomic insinuations of substituting outdated feeding approaches with automated systems (Kumar et al., 2023). In conclusion, the absence of calibration and interoperability between diverse drone models, AI platforms, and feeding apparatuses additional confounds the incorporation efforts. Overcoming these encounters needs interdisciplinary collaboration, progressions in drone battery technology, and more comprehensive policymaking that aids the adoption of AI-ambitious drone systems in agricultural and aquacultural feeding applications.

Future perspectives and research opportunities

The incorporation of Artificial Intelligence (AI) with drone technologies in the field of meticulous agriculture, livestock management, and aquaculture systems signifies a transformative leap. As AI-ambitious drones become more erudite, the future grips enormous probable for mechanizing and augmenting feeding procedures across innumerable sectors. One auspicious area is the expansion of intelligent feeding systems that apply AI algorithms to examine instantaneous environmental data, animal behavior, and nutritional necessities, thereby distributing accurate feed measures. This pointedly decreases feed leftover and rallies growth rates, particularly in remote and extensive operations (Zhang et al., 2020). In aquaculture, AI-combined drones armed with submerged radars and imaging utensils screen fish activity, biomass density, and water quality to modify feeding schedules, thus guaranteeing healthier stock and minimal environmental impact (Kim et al., 2023). Additionally, the integration of machine learning and deep learning models into drone platforms permits constant learning and adaptation to altering circumstances, permitting for prognostic analytics in feed management. This comprises predicting feed consumption patterns based on weather trends, growth stages, and behavioral indicators. Another developing frontier is the use of swarm drone technology coordinated by AI to cover wide-ranging agricultural or aquatic areas more proficiently, confirming uniform feed delivery and instantaneous monitoring (Jain et al., 2022). Merging this technology into blockchain improves the traceability and transparency of feed logistics and quality control, which is vital for regulatory compliance and customer trust. The future studies in this system will be focused on integrating multimodal sensors, edge computing, and 5G connectivity to support high-speed data processing and decision-making directly on the drone platform. This technology has some drawbacks, such as drone energy efficiency, flight autonomy, robustness of AI models under different environmental conditions, ethical and policy implications to data privacy, animal welfare and deployment. The rectification of these challenges makes these technologies more successful in future. The collaboration between AI specialists, agricultural scientists and robotic engineers will advance this technology in the feeding sector. Overall, the integration of AI and drone technologies in the agricultural and aquacultural sectors leads to profitability, sustainability and efficiency in feeding practices (Ramesh et al., 2023).

Conclusion

The AI-powered drones ensure the precise feeding for large-scale fields by real-time monitoring. Through this screening, they implement selective and targeted feeding and augment material usage. This system investigates the environmental data, feeding patterns and makes self-governing determinations, decreasing the labor-intensive and operational charges. Moreover, incorporated AI algorithms and drone mobility confirm longevity and productivity by way of evolving feeding approaches. The worldwide need toward seafood rises, and these revolutions in the food industry confirm food security and ecological conservation. The incessant research, partnership and deliberation of moral implications are crucial to improve these technologies and enhance their accountable arrangements. In the end, AI incorporation into drone technology is an auspicious step toward the forthcoming intelligent, automatic and longevity feeding approaches across various domains.

References

Balafoutis, A., Beck, B., Fountas, S., Tsiropoulos, Z., Vangeyte, J., van der Wal, T., ... Gemtos, T. A. (2017). Precision agriculture technologies positively contribute to GHG emissions mitigation, farm productivity and economics. *Sustainability, 9*(8), 1339.

Berckmans, D. (2017). General introduction to precision livestock farming. *Animal Frontiers, 7*(1), 6–11.

FAO. (2022). *The state of food and agriculture 2022: Leveraging automation in agriculture.* Food and Agriculture Organisation of the United Nations.

Ghosh, M., & Rihel, J. (2020). Hierarchical compression reveals sub-second to day-long structure in larval zebrafish behavior. *Eneuro, 7*(4).

Gutiérrez, J., Villa-Medina, J. F., Nieto-Garibay, A., & Porta-Gándara, M. A. (2021). Automated irrigation system using a wireless sensor network and a GPRS module. *IEEE Transactions on Instrumentation and Measurement, 63* (1), 166–176.

Jain, S., Verma, R., & Srivastava, M. (2022). Swarm drone technology and AI: The future of smart farming. *Journal of Agricultural Technology, 18*(4), 1345–1357.

Kamilaris, A., & Prenafeta-Boldú, F. X. (2018). Deep learning in agriculture: A survey. *Computers and Electronics in Agriculture, 147*, 70–90.

Khan, S. A., Zia-ul-haq, H. M., Umar, M., & Yu, Z. (2021). Digital technology and circular economy practices: An strategy to improve organizational performance. *Business Strategy & Development, 4*(4), 482–490.

Kim, J., Park, E., Cho, S., Kwon, K., & Ko, Y. M. (2023). Probabilistic modeling of fish growth in smart aquaculture systems. *KSII Transactions on Internet & Information Systems, 17*(8).

Kumar, N., Mallick, P. K., & Verma, S. (2019). Role of cloud-IoT in smart farming: Challenges and opportunities. In *Smart agriculture* (pp. 185–204). Springer.

Kumar, S., Verma, A. K., Singh, S. P., & Awasthi, A. (2023). Immunostimulants for shrimp aquaculture: Paving pathway towards shrimp sustainability. *Environmental Science and Pollution Research, 30*(10), 25325–25343.

Liakos, K. G., et al. (2018). Machine learning in agriculture: A review. *Sensors, 18* (8), 2674.

Liu, Y., Wang, H., Li, M., & Yang, X. (2020). Drone-based remote sensing in aquaculture: Applications and challenges. *Aquaculture International, 28*(2), 687–702.

Ramesh, P., Anbu, R., & Kumar, S. (2023). Next-generation drone technologies in AI-powered agriculture. *International Journal of Smart Agriculture, 5*(2), 78–92.

Rejeb, A., Keogh, J. G., Zailani, S., Treiblmaier, H., & Rejeb, K. (2022). Drones in agriculture: A review and bibliometric analysis. *Computers and Electronics in Agriculture, 198,* Article 107122.

Sankaran, S., Khot, L. R., Espinoza, C. Z., Jarolmasjed, S., Sathuvalli, V. R., Vandemark, G. J., Miklas, P. N., Carter, A. H., Pumphrey, M. O., Knowles, N. R., & Pavek, M. J. (2015). Low-altitude, high-resolution aerial imaging systems for row and field crop phenotyping: A review. *European Journal of Agronomy, 70,* 112–123.

Shahbazi, M., Théau, J., & Ménard, P. (2021). Recent applications of unmanned aerial imagery in precision agriculture: A review. *Remote Sensing, 13*(4), 756.

Shamshiri, R. R., Kalantari, F., Ting, K. C., Thorp, K. R., Hameed, I. A., Weltzien, C., ... Alameen, A. (2018). Advances in greenhouse automation and controlled environment agriculture: A transition to plant factories and urban agriculture. *International Journal of Agricultural and Biological Engineering, 11*(1), 1–22.

Sharma, A., Chahal, A. S., Singh, K. K., & Pathania, R. (2025). AI and drone technologies in agriculture. In Bansal N.K. & Thakur N. (Eds.), *Harmony in the fields: Your complete natural farming guide* (pp. 286–296).

Shi, B. S., Cheng, X. J., Chen, H. Z., Xie, J., Zhou, Z. H., Jiang, S. Q., Peng, X. M., Zhang, Y. D., Zhu, D. T., & Lu, Z. Y. (2022). Occurrence, source tracking and removal of antibiotics in recirculating aquaculture systems (RAS) in southern China. *Journal of Environmental Management, 324,* 116311.

Tilman, D., Balzer, C., Hill, J., & Befort, B. L. (2011). Global food demand and the sustainable intensification of agriculture. *Proceedings of the National Academy of Sciences, 108*(50), 20260–20264.

Tripicchio, P., Satler, M., Dabisias, G., Ruffaldi, E., & Avizzano, C. A. (2015). Towards smart farming and sustainable agriculture with drones. In *Proceedings of the International Conference on Intelligent Environments (IE)* (pp. 140–143).

van der Voort, S. R., Incekara, F., Wijnenga, M. M., Kapsas, G., Gahrmann, R., Schouten, J. W., Nandoe Tewarie, R., Lycklama, G. J., De Witt Hamer, P. C., Eijgelaar, R. S., & French, P. J. (2023). Combined molecular subtyping, grading, and segmentation of glioma using multi-task deep learning. *Neuro-oncology, 25*(2), 279–289.

Wang, Y., Liu, Z., Wang, S., & Chen, J. (2020). Intelligent aquaculture with machine learning and AI: A review. *IEEE Access, 8,* 23690–23712.

Wolfert, S., et al. (2017). Big data in smart farming–A review. *Agricultural Systems, 153,* 69–80.

Yang, G., Liu, J., Zhao, C., Li, Z., Huang, Y., Yu, H., & Wang, H. (2022). Agricultural remote sensing big data: Management and applications. *Computers and Electronics in Agriculture, 193,* Article 106626.

Yang, Z., Yu, C. H., & Buehler, M. J. (2021). Deep learning model to predict complex stress and strain fields in hierarchical composites. *Science Advances, 7*(15), eabd7416.

Yue, K., & Shen, Y. (2022). An overview of disruptive technologies for aquaculture. *Aquaculture and Fisheries, 7*(2), 111–120.

Zhang, C., & Kovacs, J. M. (2012). The application of small unmanned aerial systems for precision agriculture: A review. *Precision Agriculture, 13*(6), 693–712.

Zhang, C., Su, H., & Li, X. (2020). Optimisation of UAV path planning using an improved genetic algorithm. *Sensors, 20*(22), 6360.

Zhang, Y., Sun, H., Zhang, J., & Zhou, G. (2021). Application of artificial intelligence in smart agriculture: A review. *Information Processing in Agriculture, 8*(1), 1–13.

Zhou, Y., Zhang, Z., & Wang, Z. (2020). A review of the application of unmanned aerial vehicles for precision agriculture. *Computers and Electronics in Agriculture, 176,* Article 105658.

8

Enhancing aquatic animal health with AI-integrated bioengineering

Introduction

The international aquaculture manufacturing has experienced extraordinary development over the previous few years, making substantial aids to sustenance refuge, financial progress, and the enhancement of countryside source of revenue. Nevertheless, as manufacturing increases, encounters associated with aquatic organism well-being have become more definite, requiring progressive explanations that can endure manufacturing development while minimizing conservational and financial hazards. Illnesses, including viral hemorrhagic septicemia, white spot syndrome virus, and bacterial contagions by *Vibrio* species, persist as major constraints, habitually subsequent in disastrous sufferers (Austin & Austin, 2016; Lightner, 2011). AI-ambitious bio radars united with instantaneous water quality monitoring structures have empowered premature recognition and prognostic displaying of VHS epidemics, pointedly dipping transience rates in fish ranches. Moreover, machine learning algorithms are pragmatic to genomic and conservational data have enabled the expansion of resistant fish strains and enlarged vaccination manners. These united approaches have not only improved biosecurity and longevity in aquaculture but also established ascendable models for embarking upon other communicable ailments, thus strengthening the necessity for sustained multidisciplinary reconstruction. Outmoded illness management methods, which are mostly reliant on antibiotics, vaccines, and labor-intensive nursing, aspect abundant restrictions, as well as the progress of antimicrobial resistance, analytical interruptions, and conservational contamination (Defoirdt et al., 2011). Therefore, there is a demanding necessity for revolutionary inventions that participate in pioneering machineries to empower deliberate, detailed, and maintainable management

Aquaculture with Artificial Intelligence. https://doi.org/10.1016/B978-0-443-33704-8.00001-X

of aquatic organisms' well-being. The merging of artificial intelligence (AI) and bioengineering grants an archetypal alteration in rectifying these encounters. AI, encircling machine learning (ML), deep learning (DL), and computer vision technologies, gives incomparable competencies in instantaneous monitoring, premature illness discovery, prognostic analytics, and verdict provision (LeCun et al., 2015). In the intervening time, developments in bioengineering, such as genetic modification, nanotechnology, biosensor expansion, and synthetic biology, empower erudite interferences at the molecular and cellular echelons (Langer & Vacanti, 1993; Zhang et al., 2021). Through participating AI algorithms with bioengineered systems, investigators and consultants generate ground-breaking aquaculture atmospheres were constant well-being valuation, hazard forecast, and beleaguered therapeutic distribution become possible, competent, and mountable. AI-combined bioengineering empowers nonbellicose, computerized monitoring of aquatic creatures through tomography machineries, abiliment biosensors, and conservational observation systems. For example, AI-ambitious image examination can perceive primary symptoms of anxiety, grievance, or contagion by assessing behavioral patterns, color fluctuations, or irregular actions (Qin et al., 2020). Correspondingly, bioengineered nano sensors can be entrenched or fastened to aquatic creatures to uninterruptedly screen biomarkers, including glucose levels, immune rejoinder pointers, or anxiety hormones, conveying data wirelessly to AI systems for instantaneous construal and prepared generation (Duncan et al., 2020). These intellectual systems enable not only appropriate analysis but also meticulousness interferences, including the embattled issue of nanomedicine or initiation of gene-editing utensils in rejoinder to perceived irregularities. Additionally, AI-combined bioengineering is renovating illness modeling and vaccine expansion. AI algorithms can investigate enormous genomic, proteomic, and metabolomic datasets to detect innovative biomarkers and forecast pathogen progression, thus administrative the strategy of next-generation vaccines and therapeutics (Chen et al., 2021). United with artificial biology, investigators can bioengineer probiotic bacterium, germicide peptides, or even disease-resistant aquatic organisms with improved suppleness to developing pathogens (Zhao et al., 2018). This collaboration not only decreases the reliance on outmoded chemotherapeutic properties, nevertheless also line up with the goals of longevity aquaculture by abating environmental bearings. Contempt its revolutionary potential, the incorporation of AI and bioengineering in aquatic organisms' well-being airs numerous concerns. Data quality and readiness endure perilous blockages for working out dependable AI models. Moral contemplations nearby genetic manipulation and the disposition of investigation technologies must be prudently circumvented. Furthermore, the charge and complication of applying AI-integrated bioengineering explanations require cooperative exertions between scientists, representatives, industry sponsors, and local societies to guarantee impartial access and accountable invention (FAO, 2020). This chapter inspects the current progressions, probable applications, and predictions of AI-combined bioengineering in improving the well-being of aquatic organisms, contributing an inclusive overview of the scientific, technical, and ethical features of this rapidly developing sector.

Fundamentals of bioengineering in aquaculture

Bioengineering in aquaculture is an interdisciplinary sector that combines biological sciences with engineering methodologies to augment the rearing of aquatic animals for food, research, and conservation determinations. Primarily, it includes the application of biotechnology, mechanical systems, and environmental engineering to proposal and accomplish systems that promote the well-being, expansion, and longevity of aquatic creatures. Dominant rudiments of bioengineering comprise water quality management, system design (including recirculating aquaculture systems, RAS), bioprocess monitoring, illness deterrence approaches, and genetic improvement of reared species (Fig. 8.1). Water constraints, including dissolved oxygen, pH, ammonia, and temperature, must be exactly measured through engineered systems to avert anxiety and illness epidemics. Developments in radar machinery and automation have transformed these panels, contributing to instantaneous monitoring and prognostic analytics.

Furthermore, the incorporation of bioengineering with molecular biology empowers the expansion of ailment-resistant strains through discerning breeding or genetic amendment. In current years, the merging of bioengineering with artificial intelligence (AI) has unlocked revolutionary opportunities. AI algorithms can progress multifaceted datasets from aquaculture systems, forecast well-being hazards, augment feeding commands, and empower primary illness recognition, therefore developing biosecurity and animal well-being (Martínez-Porchas & Martínez-Córdova, 2012; Xu, Pan, & Sun, 2021). Maintainable bioengineering methods also accentuate minimalizing

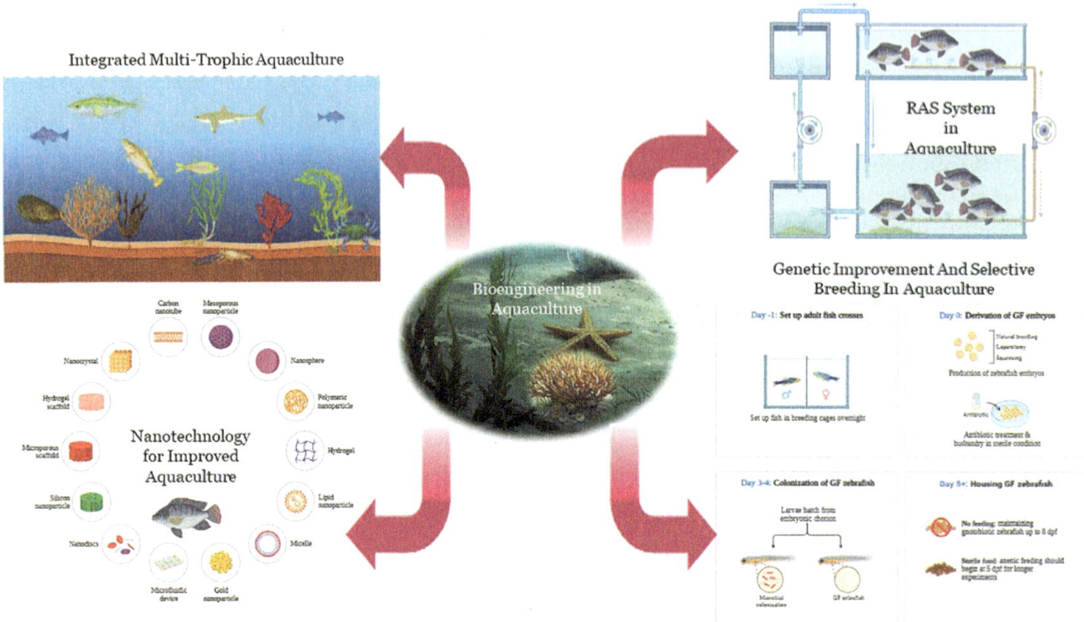

Figure 8.1 Basic fundamentals of bioengineering in aquaculture.

environmental influences through well-organized leftover management and energy preservation approaches. By way of aquaculture endures to enlarge as a momentous source of international seafood manufacture, bioengineering explanations, predominantly when combined with AI, are critical for creating buoyant and longevity aquaculture maneuvers that confirm the well-being and welfare of aquatic creatures (Tidwell & Allan, 2001; Timmons & Ebeling, 2010). Consequently, a deep sympathetic of bioengineering rudiments not only supports addressing present contests in aquatic organisms' well-being management, nonetheless also nurtures modernizations that can lead to more intellectual, adaptive, and environmentally friendly aquaculture applies.

Artificial intelligence in aquaculture: A rising frontier

Aquaculture has developed as a vital sector in certifying worldwide food safety, secretarial for virtually partial of the biosphere's fish and seafood source. By way of the manufacturing nurtures, it faces abundant contests, as well as illness epidemics, conservation vagaries, feed optimization, and wider longevity apprehensions. In current years, Artificial Intelligence (AI) has increased as a revolutionary force, contributing groundbreaking explanations to numerous of these persistent problems. AI technologies, including machine learning (ML), deep learning (DL), computer vision, and Internet of Things (IoT) incorporation, are being progressively arrayed to screen, forecast, and achieve numerous features of aquaculture processes with extraordinary exactness (Kim et al., 2022; Misimi et al., 2020). One of the furthermost substantial aids of AI in aquaculture mendacities in the province of aquatic animal well-being management. Outdated approaches of illness recognition and diagnosis, which frequently depend on labor-intensive thought and laboratory testing, are manpower-dependent, laborious, and disposed to humanoid fault. AI-grounded analytic models, predominantly those applying deep convolutional neural networks (CNNs), can hastily and precisely investigate images of fish and shellfish to recognize disease signs (Qin et al., 2020). Besides, prognostic analytics driven by AI can investigate ancient data on water quality strictures, feeding outlines, and animal behavior to predict illness epidemics formerly they arise, empowering precautionary procedures compared with over-sensitive replies (Liu et al., 2025). Yonder's well-being monitoring, AI has transformed feeding performance through the progress of ground-breaking alimentation systems. By incorporating instantaneous radar data and machine learning algorithms, these systems augment feed delivery, thus decreasing left-over, enlightening development rates, and minimalizing ecological effects, including eutrophication (Jiang et al., 2023). Computer vision applications, employing submerged cameras and image dealing out algorithms, can evaluate fish proportions, biomass, and performance, empowering accurate modifications in feeding approaches and stocking solidities. Such data-ambitious methods not only progress manufacturing competence but also endorse animal well-being by preserving ideal circumstances (Gutiérrez et al., 2018). The amalgamation of AI with bioengineering has asserted the boundaries of aquatic well-being management. Bioengineered bio radars implanted within aquaculture systems can uninterruptedly monitor biological markers symptomatic of anxiety, illness, or environmental toxicity. AI algorithms construe this multifaceted biological data, providing simultaneous verdict support to ranchers (Sundaram et al., 2020). Recent developments have seen the expansion of AI-determined autonomous robotic systems proficient in sampling, analyzing, and replying to variations within aquaculture environments without humanoid interference. These systems' aptitude is an upcoming where meticulousness in aquaculture converts the standard, categorized by negligible environmental effect and greatest animal well-being and efficiency. Though, contempt its massive possible, the application of AI in aquaculture is not without encounters. The essential for superior, characterized datasets remainders a critical blockage in training vigorous AI models. Aquaculture systems parade momentous capriciousness athwart species, geographic positions, and management applies, demanding the individualization of AI utensils to be framework-precise. Principled deliberations, counting data privacy, the supplanting of human labor, and the inadvertent environmental significances of AI-ambitious escalation, must also be prudently circumvented. Contemplating the merging of AI with developing technologies, including genomics, synthetic biology, and nanotechnology, holds marvelous capacity. AI can hasten discerning breeding programs by forecasting the finest genetic traits for illness resistance and conservation adaptation. Furthermore, AI-determined strategies of nanomaterials and embattled drug distribution schemes could transform illness treatment and vaccination approaches in aquaculture species (Pham et al., 2023). Cooperative exertions among AI experts, marine biologists, aquaculture physicians, and representatives will be vital to fully comprehend the aids of this technological uprising while safeguarding supportable and moral applies. By way of the incorporation of AI with bioengineering endures to developed, it is dignified to essentially redesign the future of aquaculture, making it more buoyant, competent, and environmentally accountable.

Integrating AI with bioengineering for disease detection and diagnosis

The amalgamation of artificial intelligence (AI) with bioengineering has unbolted novel frontiers in the primary discovery and finding of illnesses, predominantly in the sector of aquatic organisms' well-being (Fig. 8.2). Bioengineering offers groundbreaking utensils, such as bio radars, lab-on-a-chip devices, and genetically engineered biomarkers, that empower the monitoring of the functional and biochemical status of aquatic animals. Once merged with AI technologies such as machine learning (ML), deep learning (DL), and computer vision, these bioengineering utensils can distribute instantaneous, highly

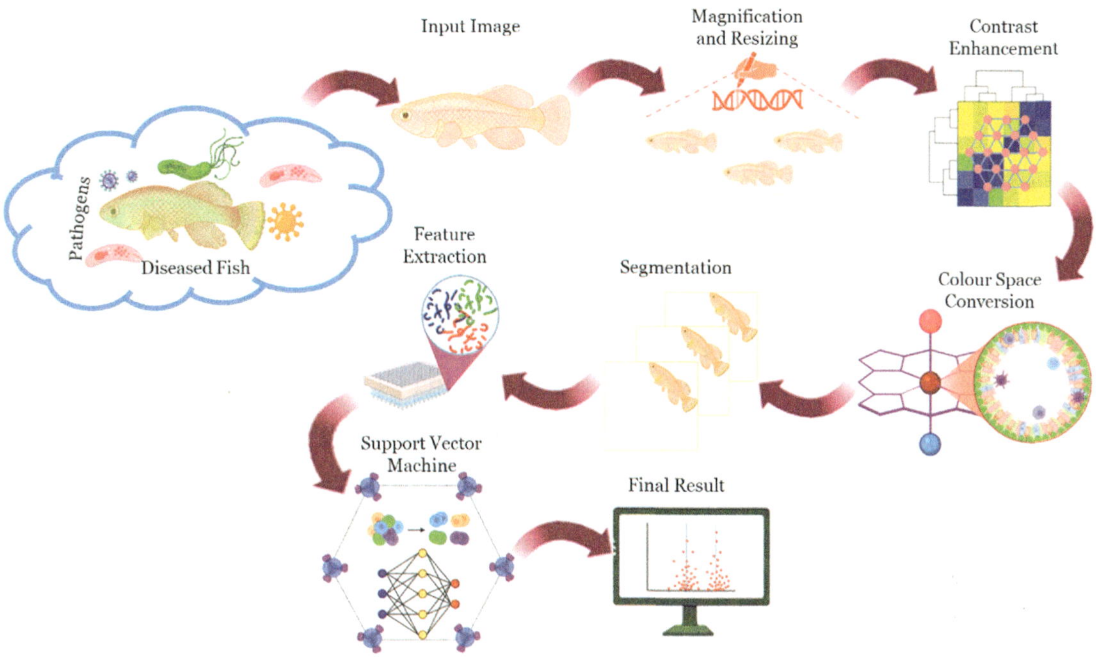

Figure 8.2 Integrating artificial intelligence (AI) with bioengineering in early detection and diagnosis of diseases.

precise illness forecasting systems. AI algorithms can investigate multifaceted biological data, such as genomics, proteomics, metabolomics, and conservational constraints, to sense delicate designs related to the commencement of illness well beforehand medical signs become observable (Esteva et al., 2017; LeCun et al., 2015). For example, computer vision systems driven by convolutional neural networks (CNNs) have been magnificently functional to recognize peripheral symptoms of contagions, including sores or fin rot in fish, with extraordinary exactness (Xu, Ma, & Zhang, 2021). Correspondingly, AI-combined bio radar areas can uninterruptedly screen water quality constraints and host physiological replies, inevitably generating signals when aberrations from usual well-being patterns are noticed (Singh et al., 2020). Additionally, AI models skilled on ancient and instantaneous data can forecast epidemics constructed on conservational stressors, permitting timely interventions to decrease indisposition and death rates (Jin et al., 2021). This collaboration between AI and bioengineering not only advances analytic haste and accuracy but also reduces anthropoid error, improves scalability, and condenses working charges, making it extremely treasured for comprehensive aquaculture maneuvers. By empowering custom-made aquatic well-being management approaches, this combined method aptitudes a architype change from sensitive to practical healthcare, safeguarding longevity and prolific aquaculture ecosystems. By way of progressions endure, the expansion of explicable AI (XAI) models and morally intended smart bio-devices will be vital for encouraging faith and extensive implementation in practical aquaculture applies (Holzinger et al., 2019).

AI-guided probiotic and vaccine development

The incorporation of artificial intelligence (AI) in the expansion of probiotics and vaccines inscriptions a revolutionary rise in improving the well-being of aquatic creatures. Outmoded methods to probiotic preparation and vaccine strategy have depended deeply on experimental approaches, which are laborious, human-powered, and habitually inadequate due to the difficulty of host-microorganism-pathogen communications. AI-ambitious systems, predominantly machine learning (ML) and deep learning (DL) algorithms, are now transforming this interplanetary by permitting hasty airing and forecasting of probiotic contenders and vaccine determinants with unparalleled exactness. AI models can examine massive datasets encircling microbial genomes, host protective responses, environmental parameters, and pathogen virulence aspects to recognize the finest probiotic strains that improve well-being, increase immunity, and progress illness resistance in aquatic animals (Fig. 8.3) (Kumar et al., 2021). Correspondingly, AI-motorized

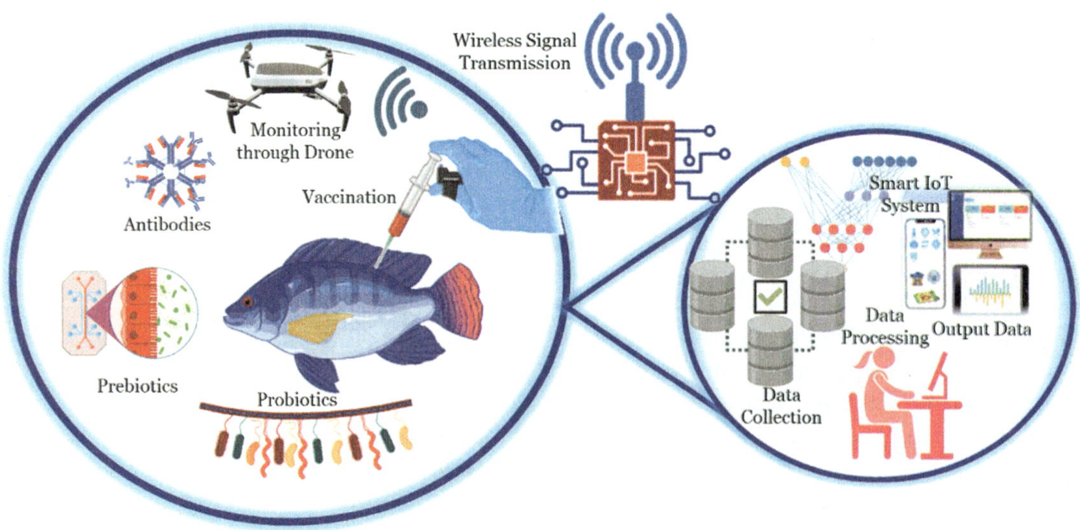

Figure 8.3 Integration of artificial intelligence (AI) in monitoring the utilization of probiotic and vaccine development in aquaculture.

immunoinformatic utensils can envisage antigenic factors, archetypal protein edifices, and pretend host-pathogen connections to design operative subunit vaccines, DNA vaccines, and mRNA vaccines precisely personalized for aquaculture organisms (Tang et al., 2025). Especially, strengthening erudition and generative adversarial networks (GANs) are being discovered to augment multiepitope vaccine paradigms by forecasting constancy, immunogenicity, and cross-protection across miscellaneous aquatic organisms' inhabitants (Zhou et al., 2025). Furthermore, AI can assist in examining vaccine effectiveness postdeployment by investigating well-being chronicles and ecological data, permitting instantaneous alterations to immunization approaches. This meticulous bioengineering method not only decreases the dependence on antibiotics but also reports developing pressures from developing pathogens and altering aquatic atmospheres. By merging AI-guided procedures, aquaculture productions can attain longevity well-being management maneuvers, confirming higher yield and biosecurity (He et al., 2023). Upcoming expansions will prospective effort on incorporating AI with high-output omics technologies, including metagenomics, transcriptomics, and proteomics, to added enhance probiotic preparations and vaccine designs for precise aquaculture environments, thus concrete the way for custom-made aquatic well-being resolutions.

Smart Aquatic Health Monitoring Systems

The rising mandate for longevity aquaculture applies has determined a revolution in aquatic organisms' well-being management, concluding in the expansion of Smart Aquatic Health Monitoring Systems (SAHMS). These systems influence progressions in artificial intelligence (AI), bioengineering, and the Internet of Things (IoT) to offer instantaneous, noninvasive scrutiny of aquatic atmospheres and organismic well-being. Contemporary SAHMS fit in radars proficient in sensing vital constraints, including water temperature, pH, dissolved oxygen, ammonia, and pathogen occurrence, thus empowering primary discovery of anxiety circumstances and illness epidemics (Martins et al., 2012; Reverter et al., 2020). AI algorithms, with machine learning models, progress the complex datasets gathered, recognizing patterns that come first noticeable indications of well-being corrosion, hence permitting for appropriate intrusion. Bioengineered abiliment or implantable bio radars have added boosted system compassion, conferring physiological perceptions at the specific visceral level, together with heart rate, respiration rate, and metabolic biomarkers (Rogers et al., 2019). Remote monitoring platforms armed with prognostic analytics not only decrease physical workers but also improve alimentation timetables, water exchange rates, and medications, enlightening mutually animal wellbeing and ranch productivity. Furthermore, combined decision-support systems can endorse educative activities based on prognostic hazard valuations, evolving meticulousness aquaculture performs (Neethiselvan et al., 2023).Contemporary innovations also prominence on energy-autonomous devices that apply bacterial fuel cells or solar energy, thus improving system sustainability in remote or seaward ranches (Park et al., 2018). By merging AI, bioengineering, and sustainable energy explanations, SAHMS epitomizes a revolutionary rise to buoyant aquaculture environments. They represent an archetype alteration from mercurial to upbeat aquatic well-being management, eventually contributing to worldwide efforts in food security and ecological preservation. By way of technology develops, sustained multidisciplinary association among marine biologists, engineers, data scientists, and aquaculture physicians will be critical to decontaminating system competencies and confirming their extensive adoption.

Precision feeding and water quality management using AI models

Meticulousness, feeding, and water quality management are vital characteristics of contemporary aquaculture expected to improve aquatic organisms' well-being, refining manufacturing competence, and minimalizing environmental effects. Outmoded approaches to handling alimentation and water quality frequently depend on physical annotations and secure timetables, which can result in overfeeding, underfeeding, or worsening of water quality parameters. Current developments in artificial intelligence (AI) models, predominantly machine learning (ML) and deep learning (DL) methods, have transformed these performances by permitting instantaneous, data-intensive decision-making. AI-combined meticulousness feeding schemes examine constraints, including fish biomass, metabolic rates, behavior patterns, and ecological circumstances (e.g., temperature, dissolved oxygen, and ammonia levels) to augment food distribution in amount, timing, and alignment (Xie et al., 2021). Computer vision algorithms, applying submerged cameras and radars, screen fish craving and commotion, confirming that feed conversion ratios are augmented while minimalizing leftover (Shi et al., 2020). Correspondingly, AI-determined water quality management systems utilize Internet of Things (IoT) radar networks to uninterruptedly collect data on pH, salinity, turbidity, and toxic constituents. Prognostic models can sense delicate leanings demonstrating probable water quality corrosion previously it develops dangerously, permitting for defensive activities such as aeration fine-tuning, filtration, or water exchange (Wang et al., 2022). Additionally, merging AI with bioengineering revolutions, including smart bioreactors and automated filtration units, improves the flexibility of aquaculture systems against illness epidemics related to poor water quality. Underpinning learning models have been applied to advance adaptive approaches that not only preserve the finest environmental circumstances but also respond enthusiastically to unexpected fluctuations triggered by peripheral stressors (Li et al., 2023). These machineries decrease the dependence on antibiotics and chemical medications, encouraging a more longevity and well-being-engrossed aquaculture practice. Inclusive, the amalgamation of AI in meticulousness feeding and water quality management characterizes a substantial rise toward attaining meticulousness aquaculture, with extensive advantages for the wellbeing of aquatic creatures, ranch productivity, and conservational preservation.

Ethical, regulatory, and economic considerations

The amalgamation of artificial intelligence (AI) and bioengineering in improving the well-being of aquatic creatures offers abundant prospects. Motionless, it also increases substantial ethical, regulatory, and economic encounters that must be prudently circumvented. Ethically, the practice of AI-ambitious machineries for nursing, analyzing, and treating aquatic animals needs an assurance to animal well-being moralities. There is a rising apprehension that overdependence on automated systems could result in reification in animal care, endangering the abandonment of separate well-being gradations (Bryson, 2018). Additionally, the gathering, storage, and consumption of enormous datasets associated to the well-being of aquatic organisms increase problems regarding data privacy, ownership, and probable misappropriation (Floridi, 2018). Supervisory frameworks must progress to retain stride with these inventions, guaranteeing that AI-grounded bioengineering utensils conform with prevailing ethics for animal well-being and biosafety standards. Present guidelines, intended mostly for conservative biotechnology and veterinary interferences, may not sufficiently discourse the dynamic and adaptive nature of AI algorithms (European Commission, 2021). Consequently, apprizing strategies to comprise strategies for algorithm transparency, accountability, and postdeployment monitoring is vital. Frugally, the early charges of emerging and organizing AI-combined bioengineering explanations can be momentous, possibly making fences for small- and medium-scale aquaculture manufacturing. Nevertheless, in the long term, these technologies aptitude augmented productivity, condensed illness epidemics, and lesser humanity rates, resulting to considerable economic benefits. Planned reserves, public-private partnerships, and subsidized access to technology could help democratize the practice of these progressive utensils. Prominently, moral and financial deliberations must be combined into supervisory decision-making to guarantee that technological advancements in aquatic well-being management promote maintainable, reasonable, and accountable development across the worldwide aquaculture manufacturing.

Future perspectives

The incorporation of artificial intelligence (AI) into bioengineering for improving the well-being of aquatic creatures aims to transform the technique we method the management and medication of aquatic organisms in mutually cultivated and wild atmospheres. By way of the worldwide request for aquatic diets is expected to continue to rise, sustaining the well-being of aquatic creatures is vital for confirming longevity, manufacturing and environmental equilibrium. AI technologies, together with machine learning, computer vision, and prognostic analytics, provide unique prospects for enlightening illness finding, inhibition, and medication, thus evolving bioengineering approaches for aquatic organisms' well-being. In the forthcoming, AI-empowered analytic utensils are predictable to be combined into instant monitoring systems, allowing the initial discovery of illnesses through the investigation of behavioral patterns, environmental circumstances, and physiological fluctuations in aquatic creatures. For example, machine learning algorithms can progress hefty datasets from radars and cameras to recognize elusive symptoms of anxiety, contagion, or other well-being apprehensions beforehand they become noticeable to human spectators. This initial recognition competence will enable the appropriate interference and decrease the blowout of illnesses, eventually resulting in improved aquaculture processes and more sustainable environments. Additionally, AI can assist in the expansion of bioengineered explanations, including genetically modified organisms (GMOs) or meticulousness medication methods, that augment ailment resistance in aquatic animals, resulting in amended development rates and enlarged confrontation to pathogens. Added key expansion is the advantage of AI in augmenting aquatic ecosystems. Bioengineering explanations combined with AI can aid in preserving the best circumstances for aquatic animals by incessantly screening water quality, temperature, salinity, and other vital constraints simultaneously. AI-determined systems can not only screen these variables but also forecast and alleviate variations that may pressure aquatic creatures or result in illness epidemics. For instance, prognostic analytics can predict destructive algal blooms, permitting initial countermeasures to defend aquatic animals from toxins. Moreover, AI could improve the strategy of bioengineering involvement, including further competent water filtration systems or automated feeding technologies, confirming that these systems are adaptable and proficient in retorting to the dynamic requirements of aquatic environments. Furthermore, AI-combined bioengineering extends massive possibilities for custom-made medication in aquatic organisms' well-being. Through examining genomic, conservational, and well-being data from distinct animals, AI algorithms could offer perceptions into the precise requirements of dissimilar species or even specific animals, permitting the expansion of custom-made conducts and preemptive procedures. This meticulousness method could improve the effectiveness of vaccination curricula, enhance alimentary approaches, and decrease the practice of antibiotics and other chemicals, thus minimalizing environmental impressions and improving animal well-being. In relation to investigation, the forthcoming will expected see a merging of AI and bioengineering in the examination of innovative therapeutics, including biomaterials and bioactive composites, consequent from aquatic animals themselves.

AI-driven imitations and bioinformatics utensils could accelerate the detection of these complexes and their possible advantages in averting and treating illnesses in aquatic creatures. Furthermore, the amalgamation of AI and bioengineering could result in innovations in the expansion of more buoyant aquaculture systems that can endure the encounters postured by climate alteration, such as increasing ocean temperatures and transformed disease dynamics. Contempt the auspicious stance, numerous contests must be addressed to fully comprehend the probable of AI-combined bioengineering in aquatic organisms' well-being. Moral apprehensions concerning genetic modifications, data privacy, and the ecological impacts of prevalent technological advantages must be prudently measured. As well, the effective distribution of AI explanations in aquatic creatures' well-being will necessitate multidisciplinary association between bioengineers, data scientists, ecologists, and veterinarians. As the technology endures to progress, it will be vital to begin consistent etiquettes and governing agendas to confirm the harmless and accountable advantages of AI in aquaculture and marine biology.

Conclusion

AI-determined technologies, including machine learning algorithms, prognostic analytics, and instantaneous screening systems, permit investigators and physicians to report the well-being encounters met by aquatic animals proactively. The application of AI in bioengineering permits initial disease recognition, custom-made medication approaches, and the effective management of aquatic environments, thus encouraging enhanced longevity in both aquaculture and wild populations. Additionally, the meticulousness accessible by AI confirms a more besieged approach to handling pathogens, ecological stressors, and genetic syndromes. By way of AI endures to progress, its incorporation with bioengineering is probable to transform the sector of aquatic organisms' well-being, offering new prospects for invention, scalability, and augmented resilience within aquatic systems. Through sustained multidisciplinary research and alliance, AI can make a momentous involvement to the expansion of sustainable, well-organized, and vigorous practices for the forthcoming of aquatic organisms care.

References

Austin, B., & Austin, D. A. (2016). *Bacterial fish pathogens: Disease of farmed and wild fish*. Springer.

Bryson, J. J. (2018). Patiency is not a virtue: The design of intelligent systems and systems of ethics. *Ethics and Information Technology, 20*(1), 15–26.

Bürgin, A. (2021). The European Commission. In *Environmental policy in the EU* (pp. 93–109). Routledge.

Chen, R., et al. (2021). Artificial intelligence in aquaculture: Applications, challenges, and future perspectives. *Aquaculture International, 29*(2), 679–700.

Defoirdt, T., et al. (2011). Alternatives to antibiotics for the control of bacterial disease in aquaculture. *Current Opinion in Microbiology, 14*(3), 251–258.

Duncan, B., et al. (2020). Smart nanosensors for monitoring aquatic health. *Sensors and Actuators B: Chemical, 320,* Article 128388.
Esteva, A., Kuprel, B., Novoa, R. A., et al. (2017). Dermatologist-level classification of skin cancer with deep neural networks. *Nature, 542*(7639), 115–118.
FAO. (2020). *The state of world fisheries and aquaculture 2020* (pp. 1–244). Food and Agriculture Organization of the United Nations.
Floridi, L. (2018). Soft ethics and the governance of the digital. *Philosophy & Technology, 31*(1), 1–8.
Gutiérrez, M., Biagioni, R. N., Alarcón-Herrera, M. T., & Rivas-Lucero, B. A. (2018). An overview of nitrate sources and operating processes in arid and semiarid aquifer systems. *Science of the Total Environment, 624,* 1513–1522.
He, X., Abakari, G., Tan, H., Liu, W., & Luo, G. (2023). Effects of different probiotics (*Bacillus subtilis*) addition strategies on a culture of *Litopenaeus vannamei* in biofloc technology (BFT) aquaculture system. *Aquaculture, 566,* 739216.
Holzinger, A., Biemann, C., Pattichis, C. S., & Kell, D. B. (2019). What do we need to build explainable AI systems for the medical domain? *Review and recommendations. Artificial Intelligence in Medicine, 98,* 3–13.
Jiang, L., Lu, X., Wang, G., Peng, M., Wei, A., Zhao, Y., & Soetaert, K. (2023). Unraveling seasonal and interannual nutrient variability shows exceptionally high human impact in eutrophic coastal waters. *Limnology and Oceanography, 68*(5), 1161–1171.
Jin, Y., Wang, Y., Zhang, Z., & Zhang, Z. (2021). Application of machine learning for aquatic disease prediction: A review. *Aquaculture Research, 52*(12), 5994–6007.
Kim, J., Kim, Y., Park, S. E., Kim, T. H., Kim, B. G., Kang, D. J., & Rho, T. (2022). Impact of aquaculture on distribution of dissolved organic matter in coastal Jeju Island, Korea, based on absorption and fluorescence spectroscopy. *Environmental Science and Pollution Research, 29*(1), 553–563.
Kumar, V., Roy, S., Behera, B. K., Bossier, P., & Das, B. K. (2021). Acute hepatopancreatic necrosis disease (AHPND): Virulence, pathogenesis and mitigation strategies in shrimp aquaculture. *Toxins, 13*(8), 524.
Langer, R., & Vacanti, J. P. (1993). Tissue engineering. *Science, 260*(5110), 920–926.
LeCun, Y., Bengio, Y., & Hinton, G. (2015). Deep learning. *Nature, 521*(7553), 436–444.
Li, H., Cui, Z., Cui, H., Bai, Y., Yin, Z., & Qu, K. (2023). A review of influencing factors on a recirculating aquaculture system: Environmental conditions, feeding strategies, and disinfection methods. *Journal of the World Aquaculture Society, 54*(3), 566–602.
Lightner, D. V. (2011). Virus diseases of farmed shrimp in the Western hemisphere (the Americas): A review. *Journal of Invertebrate Pathology, 106*(1), 110–130.
Liu, Y., Deng, Y., Wu, Q., Jin, C., Zhao, Y., Gao, M., & Guo, L. (2025). Mariculture solid waste application for marine recirculating aquaculture system wastewater treatment: The role of neglected sulfide electron donor. *Journal of Cleaner Production, 486,* 144493.
Martínez-Porchas, M., & Martínez-Córdova, L. R. (2012). World aquaculture: Environmental impacts and troubleshooting alternatives. *The Scientific World Journal, 2012,* Article 389623.
Martins, C. I. M., et al. (2012). New developments in fish welfare indicators. *Fish Physiology and Biochemistry, 38*(1), 17–41.
Misimi, F., Kajtna, T., Misimi, S., & Kapus, J. (2020). Development and validity of the fear of water assessment questionnaire. *Frontiers in Psychology, 11,* 969.
Neethiselvan, N., et al. (2023). Smart aquaculture: Advances and challenges. *Aquaculture Reports, 27,* Article 101379.

Park, J., et al. (2018). Microbial fuel cells as a power source for aquatic biosensors. *Biosensors and Bioelectronics, 122,* 107–115.

Pham, H. D., Siddik, M. A., Rahman, M. A., Huynh, L. T., Nahar, A., & Vatsos, I. N. (2023). Effects of n-3 HUFA-enriched Artemia on growth, biochemical response, skeletal morphology and stress resistance of Asian sea bass (*Lates calcarifer*) larvae reared at high temperature. *Aquaculture, 574,* 739732.

Qin, H., et al. (2020). Machine learning in fish behavior analysis: Review and perspectives. *Reviews in Aquaculture, 12*(4), 2447–2462.

Reverter, M., et al. (2020). Use of sensors and AI for fish health and welfare monitoring. *Reviews in Aquaculture, 12*(2), 924–944.

Rogers, J. A., et al. (2019). Bio-integrated and bio-inspired materials in aquatic biosensors. *Nature Reviews Materials, 4*(8), 528–541.

Shi, F., Qiu, X., Nie, L., Hu, L., Babu, S., Lin, Q., Zhang, Y., Chen, L., Li, J., Lin, L., & Qin, Z. (2020). Effects of oligochitosan on the growth, immune responses and gut microbes of tilapia (*Oreochromis niloticus*). *Fish & Shellfish Immunology, 106,* 563–573.

Singh, V., Rai, P., & Thakur, A. (2020). Biosensors for aquatic health monitoring: Recent advances and future perspectives. *Biosensors and Bioelectronics, 165,* Article 112404.

Sundaram, T., Indu, B., Srinivasulu Reddy, C. H., Swathi Raj, V. S., Hari priya, S., Poojitha, P., Renusree, K., Govindarajan, G., Rajendiran, N., & Sundaram, V. (2020). Bio inspired silver nanoparticle synthesis from fish liver oil and its antibacterial activity against shrimp pathogen. *InIOP Conference Series: Materials Science and Engineering, 993*(1), 012166.

Tang, M., Feng, J., Wang, X., & Ding, Y. (2025). An attenuated live strain of HY9901 mutant Δgr provides protection against *Vibrio alginolyticus* in pearl gentian grouper (♀ *Epinephelus fuscoguttatus* × ♂ *Epinephelus lanceolatu*). *Veterinary Immunology and Immunopathology, 280,* 110887.

Tidwell, J. H., & Allan, G. L. (2001). Fish as food: Aquaculture's contribution. *EMBO Reports, 2*(11), 958–963.

Timmons, M. B., & Ebeling, J. M. (2010). *Recirculating aquaculture* (3rd ed.). Ithaca Publishing Company.

Wang, C., Jiang, C., Gao, T., Peng, X., Ma, S., Sun, Q., Xia, B., Xie, X., Bai, Z., Xu, S., & Zhuang, X. (2022). Improvement of fish production and water quality in a recirculating aquaculture pond enhanced with bacteria-microalgae association. *Aquaculture, 547,* 737420.

Xie, S., Wei, D., Tian, L., & Liu, Y. (2021). Dietary supplementation of chenodeoxycholic acid improved the growth performance, immune response and intestinal health of juvenile *Penaeus monodon* fed a low fish-meal diet. *Aquaculture Reports, 20,* 100773.

Xu, D., Ma, X., & Zhang, Y. (2021). Automated recognition of fish diseases using deep learning techniques. *Aquaculture, 532,* Article 735992.

Xu, L., Pan, L., Sun, X., & Liu, Y. (2021). Applications of artificial intelligence in aquaculture: Current status and future directions. *Aquaculture, 531,* Article 735886.

Zhang, Q., et al. (2021). Synthetic biology meets aquaculture: Engineering probiotics and beyond. *Trends in Biotechnology, 39*(7), 695–705.

Zhao, X., et al. (2018). Application of bioengineering in aquaculture: Opportunities and challenges. *Biotechnology Advances, 36*(4), 837–853.

Zhou, C., Wang, C., Sun, D., Hu, J., & Ye, H. (2025). An automated lightweight approach for detecting dead fish in a recirculating aquaculture system. *Aquaculture, 594,* 741433.

9

Handling IoT toward the aquaculture

Introduction

The Internet of Things (IoT) signifies a ground breaking move in data capturing, sharing and analysis in different field, which includes aquaculture. The blend of innovative sensor devices and wireless communication systems and cloud computing provides unique prospects for idealizing aquaculture methods. The fast growth of global aquaculture production field is responsible for almost half of all the consumed wish worldwide, the requirement for innovative techniques to guarantee sustainability, improves yield and reduce environmental impact. Aquaculture, frequently denoted as "fish farming" has become an important resource of protein, offering livelihoods to millions of people and significantly contributing to global food security (FAO, 2020). Though, the segment undergoes obstacles such as water quality management, disease control, environmental observations and operational competency. The IoT is rising as a crucial to result to discourse these obstacles by aiding instantaneous observation, mechanization and data-aided decision making. IoT supported devices in aquaculture, extending from sensor devices for temperature, pH, dissolved oxygen and salinity to mechanized feeding systems and fish trailing tools, provides huge data that boosts functional efficacy and progresses the well-being of farmed species. Through the incorporation of IoT systems, aquaculture farms observes environmental limitations in real-time, detect disease outbursts earlier, self-systemize feeding to decrease waste and ideal resource utilization (Zhou et al., 2021). The usage of IoT not only expands efficiency but also aids sustainable approaches by dropping water and energy ingestion, diminishing the utilization of chemicals and avoiding overfeeding all of which subsidize reduction of environmental footprint of aquaculture (Li et al., 2022). A vital feature of IoT in aquaculture is its capacity to afford nonstop, precise, and specific observation. In old-style aquaculture, labor-intensive examinations and

Aquaculture with Artificial Intelligence. https://doi.org/10.1016/B978-0-443-33704-8.00004-5

uneven sampling frequently upshot deferred identification of problems, such as water quality deprivation or disease occurrences. With IoT, farmers obtain instant warnings and crucial perceptions through interlinked devices and platforms, which permits more practical administration (Lee et al., 2020). For instance, IoT sensing devices uninterruptedly monitors water quality to follow the health and growth rates of fish, where mechanized systems alter feeding rates based on instantaneous data.

In addition, the IoT's capability to observe the site and behavior of fish over GPS and audio sensor devices has improved fish farming methods by aiding farmers to administer stock density and evade congestion, which leads to disease outbursts and decreased growth (Patel & Kumar, 2021). The role of IoT in improving aquaculture is additionally reinforced by progressions in machine learning and artificial intelligence (AI). AI algorithms analyze the huge amount of data produced by IoT devices, offering visions into outlines that may not be proximately clear to human users. This data-aided method has the possibility to suggestively augment decision-making processes, from envisaging ideal harvest times to recognizing the most efficient feeding timescales. With the combination of IoT and AI, aquaculture systems are developing near completely computerized, self-governing functions that need negligible human-involvement, thus enlightening working effectiveness and falling labor costs (Zhang et al., 2022). In spite of the capable reimbursements, the acceptance of IoT in aquaculture faces several obstacles. The cost of IoT organization, which includes sensor devices, communication webs and data storage is unaffordable particularly for small-scale fish farmers (Vijayan et al., 2020).

Moreover, the dependability of IoT systems in severe aquatic circumstances, where devices are unprotected from water, humidity and biofouling remnants as an alarm. These mechanical tasks require the progress of more vigorous and profitable IoT results specified to the exclusive circumstances of aquaculture. In addition, problems related to data safety and secrecy, especially with the growing amount of data obtained by IoT systems, necessitate deliberation and guidelines to guard complex information (Muthukumar et al., 2023).

Fundamentals of IoT architecture

The internet of Things (IoT) architecture acts as the introductory outline that empowers unified connectivity and interface between devices, sensors and applications, enabling at present

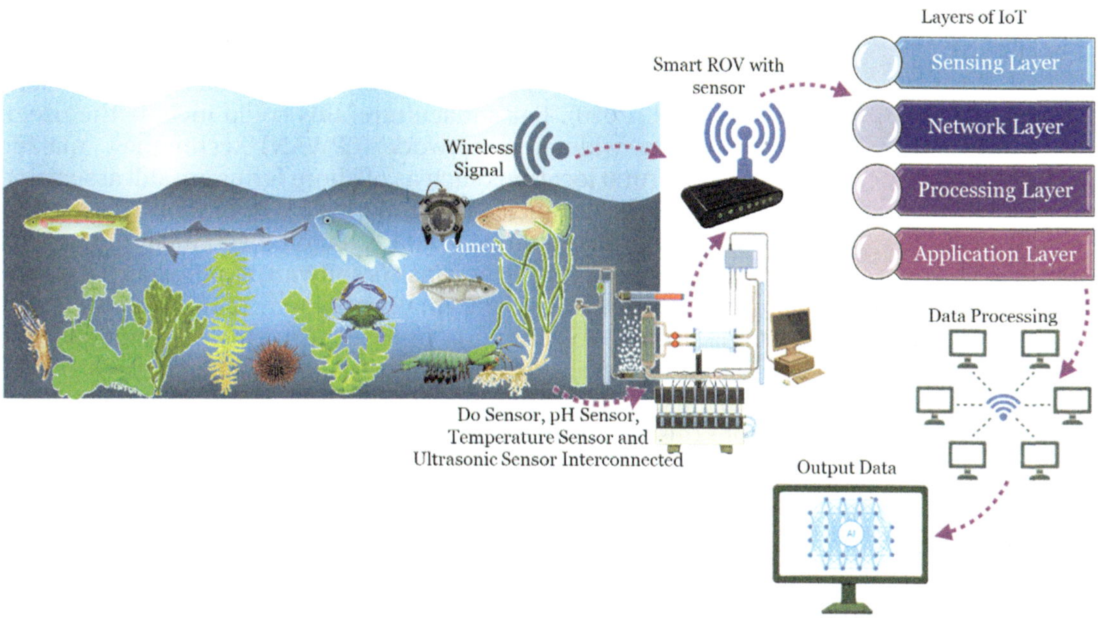

Figure 9.1 Internet of things (IoT) architecture framework enabling interaction between devices, sensors, and applications, facilitating real-time data collection, processing, and decision-making.

data collection, processing and conversant decision-making (Fig. 9.1). At its center, IoT construction is planned to sustenance a web of interrelated corporal matters that connect with one another through the internet. These devices, naturally armed with devices such as sensors and actuators, obtain data from their background and transfer it to the cloud or power systems for investigation, storage and dealing out. The IoT architecture is usually designed into three important layers such as perception, network and application layers (Gubbi et al., 2013). The perception layer is accountable for detecting the corporeal world through several strategies including sensors and actuators. These devices sense limits such as temperature, humidity, pH levels, motion, and others, which are crucial for fields such as aquaculture (Zhang et al., 2021). The sensors change physical data into digital inputs and signals, which in turn transferred to the adjacent layer for analysis. In an aquaculture system, for instance, water quality limits such as dissolved oxygen and salinity are regularly observed, to make sure the environment stays ideal for aquatic organisms.

The web system grips the broadcast of data between amongst devices and the cloud or local servers. It includes numerous communication procedures and techniques, which includes

Wi-Fi, Bluetooth, Zigbee, LoRaWAN, and cellular networks. This layer guarantees that data from sensor devices attains the dispensation and packing systems dependably and steadily (Tao et al., 2020). For aquaculture, this could include the use of low-power wide-area network (LPWAN) techniques, making sure that rural locations such as offshore farms are still associated in spite of the distance and simulating environment. The application layer is where the IoT data is handled and analyzed to offer crucial visions. This layer consists software applications, data analytics tools and decision-making system that employ the obtained data to support precise use. In aquaculture, the application layer has observing systems that track water conditions, fish health and development rates. The visions obtained from this data helps in idealizing feed methods, preventing disease outbursts and improve overall administration approaches.

Besides, it empowers instantaneous decision-making such as mechanically adjustable aeration unit depending up on oxygen levels. Among the rising progressions in IoT construction is the inclusion of superior computing. Such computing brings dispensation power quicker to the data basis, decreasing dormancy and bandwidth necessities by executing original data processing at or closer to the sensor level before transferring it to the cloud (Duan et al., 2020). This is especially useful for aquaculture systems where at present data processing is important for regulating the well-being of aquatic life. For example, in an aquaculture farm, superior devices process sensor information close by to induct instant movements such as activating water pumps or adjusting temperature levels, before distributing combined data to the cloud for further investigation and handling. Safe keeping is an important aspect of IoT systems. Given the huge amount of delicate data obtained and distributed crossways to the IoT ecosystem, safeguarding communication channels and defending devices from cyber threats are indispensable. Verification, encoding and admittance controller are essential to protect data truthfulness and avoid illegal admittance. In aquaculture systems, confirming the safety of data from water quality sensors to feed mechanisms is critical to evade interfering, which could end up in huge economic losses or damage to the environment (Abdulkareem et al., 2022).

Key IoT devices and sensors in aquaculture

In aquaculture domain, the incorporation of Internet of Things (IoT) devices and sensors has transformed out dated farming methods, permitting immediate observation, data collection and idealization of numerous operative aspects. These IoT devices are crucial in enhancing the efficacy, sustainability and cost-effectiveness of aquaculture procedures. Among them key IoT devices are water quality analyzer, environmental observation system, feed management sensors and biometric sensors for nursing fish health (Fig. 9.2). Water quality sensors are the crucial IoT devices in aquaculture, as they uninterruptedly observe critical limits such as pH, temperature, dissolved oxygen, salinity and turbidity. Regulating ideal water quality is important for the well-being and development of aquatic species. Instruments such as YSI ProDSS (Digital Sampling System) and the Hanna Instruments HI98193 multiparameter tester are widely used in aquaculture farms to identify alterations in water conditions, varying mechanists to do needed alterations to avoid fish stress or mortality. These sensors are typically united with mechanized systems to persuade activities such as ventilation or water change dependent on the immediate data. Environmental nursing systems counting climate devices, imaging device and

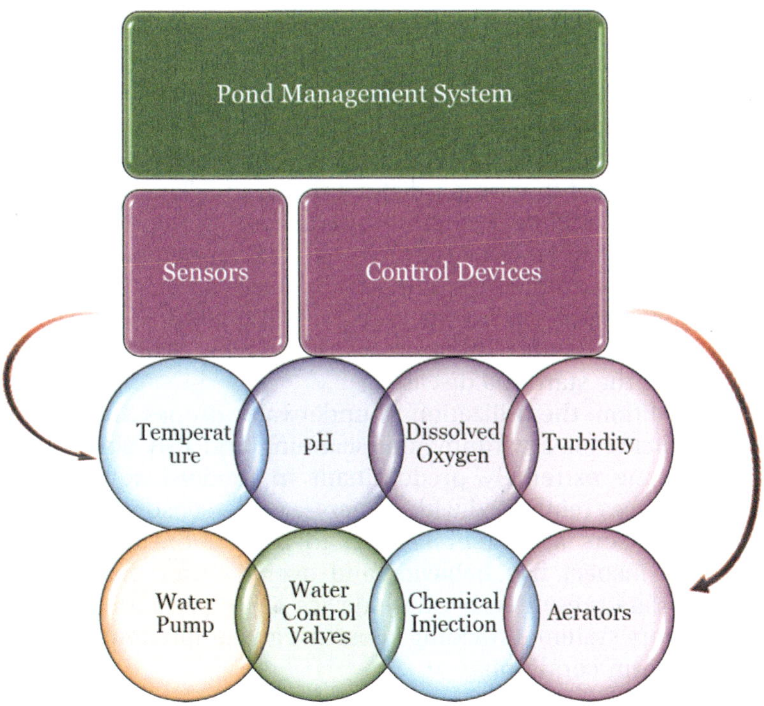

Figure 9.2 Details on key IoT devices and sensors in aquaculture.

GPS devices also has a main part in aquaculture. These devices provide real time information on marginal features such as climate, atmospheric pressure, wind speed, and sunlight strength, which interrupts water temperature, dissolved oxygen levels and overall fish well-being. Environmental data is vital for comprehensive casting and dropping ecological stressors, which leads to fish disease explosions or deprived evolving conditions. IoT-based meteorological stations such as Davis instruments afford joint outcomes to observe plentiful environmental variables contributing to a more extensive considering the aquaculture ecosystem. Feed handling in aquaculture is an extent where IoT sensors fallout in a substantial impact. IoT-armed feeding systems, such as mechanical feeders reinforced by weight sensors and cameras supports in amendable feed delivery. These instruments make sure supreme feeding approaches, thereby lessening overfeeding or underfeeding leading to water adulteration and unproductive reserve use. Innovative feeding systems such as Skretting and Innovasea utilizes sensors to control the amount of feed required created on instant data on fish health and environmental circumstances. These systems also perceive the feed conversion ratio, that is crucial for assessing the sustainability and effectiveness of fish farming activities.

Biometric instruments for fish welfare maintenance are also fast in union with aquaculture. These sensors enumerate vigorous indications such as heart rate, swimming behavior and skin settings that designates fish health status. Actions such as RFID cataloging coupled with temperature and motion sensors are used to monitor distinct fishes and classify any aberrations that could be a symptom of disease eruptions or stress. For example, RFID-based structures have been used to track fish growth and notice gestures of diseases such as gill rot and white spot syndrome (Tao et al., 2019). Mobile sensors such as accelerometers classify irregular whirling forms that designates distress or the start of a disease.

In addition, the utilization of underwater drones and motorized vehicles for environmental scanning and fish observation has become extremely predominant in modern aquaculture. These devices reinforced with cameras, audio sensors and water quality sensors are latent enough to revolve the aquaculture farms to inspect fish behavior and overall water quality. The prompt data presented by these device is merged into a central observatory system, providing operators an comprehensive valuation of farm conditions.

Network infrastructure and data communication

The expansion of vigorous network system and data communication systems is necessary to the fruitful positioning of Internet of Things (IoT) uses in aquaculture. In an IoT-supported aquaculture environment, a variety of sensors, devices and systems work organized to obtain and transfer instant data on water quality, fish health, feeding patterns and environmental situations (Siti et al., 2020). These IoT sensors, which include temperature sensors, pH sensors, and motion trackers, must be interlinked effectively and dependably to transfer data to a centralized monitoring system or cloud storage for examination and conversant decision-making. A crucial perspective of network construction in aquaculture IoT application is making sure low inactivity, more consistency and scalability, which allows nonstop observation and administration of aquaculture farms, even in rural and extreme environments (Jiang et al., 2021).

The web infrastructure aiding IoT in aquaculture generally includes the amalgamation of wired and wireless communication techniques. For local area connectivity within a farm, strategies such as Zigbee, LoRa(Long Range) and Bluetooth low Energy (BLE) are widely used for their less power utilization and range capacities (Oliveira et al., 2019). These wireless procedures are compatible for positioning dispersed sensor webs in aquatic environments where wired systems is an obstacle. For wider communication and connection with cloud platforms or rural servers, cellular webs (e.g., 4G, 5G) or satellite communication techniques are usually active, especially for aquaculture farms located in seaward or rural locations where terrestrial internet facility is less or unavailable (Ali et al., 2020). Data communication in IoT-aided aquaculture includes various steps, initiating from data collection at the sensor level to transferring them over web and final assessment. The obtained sensor data is generally transferred in instant to superior devices or gateways that assess and separate the data before forwarding it to cloud or central servers for future analysis. This superior computing method helps in decreasing the amount of firsthand data transferred over the web, thus idealizing bandwidth and reducing inexpression (Raza et al., 2017).

Additionally, approaches such as MQTT (Message Queuing Telemetry Transport) and CoAP (Constrained Application Protocol) are appointed for their nature in less-bandwidth, high-latency or unreliable system conditions, which are general in rural aquaculture farms (Hossain et al., 2018). Besides, ensuring the data protection and confidentiality is of superlatively vital in IoT aided aquaculture systems. With the growing number of intertwined devices and subtle nature of data associated to farm activities, safeguarding announcement channels and shielding against fake threats are decisive (Ganguly et al., 2020). Technical options such as endwise encryption, safe communication procedures and verification systems are hired to defend the data openness and confidentiality along the IoT network.

Data analytics and decision support systems

In regard of modern aquaculture, the inclusion of IoT techniques has transformed the way data is attained, examined and applied to earlier decision-making. The role of data analytics (DA) and Decision Support System (DSS) is essential in improving aquaculture operations, warranting sustainability and enhancing operational efficiency. IoT-aided devices in aquaculture offers delivers data on current water quality, fish health and environmental situations. These devices, such as sensors for temperature, pH levels, salinity, dissolved oxygen and turbidity collect huge amount of data that when assessed and examined, allows the idealization of aquaculture system management. Data analytics in IoT supported aquaculture system concentrates in mining expressive visions from the nonstop stream of data obtained by sensors. Through different methods, which include various projective analytics, machine learning, and statistical examination, aqua culturists can broad cast events such as disease outbursts, oxygen exhaustion, or possible developmental problems, permitting preventive actions. For instance, machine learning models envisage the growth rate of fish depending up on environmental aspects and feeding timetables. At the same time. Abnormality detection procedures classify indiscretions that may specify health issues or system faults (Yuan et al., 2022).

Also, immense data tools are hired to progress huge datasets, empowering compound investigation such as detecting trends over extended periods and across several farm areas (Zhang et al., 2022). Decision Support System (DSS) additionally boost the capability to interpret raw information into practical decisions. DSS in aquaculture integrates several data resources, which include environmental conditions, market trends, yield rates, and uses decision models to help aquaculture managers in making precautionary decisions. For instance, when water quality parameters are outside the ideal limits, a DSS will suggest remedial approaches such as adjusting aeration or adapting feed circulation.

Moreover, DSS supports tactical decision-making procedures such as site selection for new farms, resource allotment and risk administration by combining multidisciplinary data and providing scenario analysis tools (Hoch et al., 2023). Among the key benefits of operating DA and DSS in IoT-aided aquaculture, where data-aided visions allow extremely specified interferences. Unceasingly observing environmental situations, fish behavior and well-being, aquaculture processes reduce resource waste, decrease effects on environment and increases yield. Furthermore, the active nature of aquaculture systems, inclined by aspects such as climate change, alterations in water quality, and disease burden, requires fruitful decision-making. Data-driven DSS offer immediate compliance, making sure that aquaculture systems continues to be robust toward changes in both environmental and operative encounters.

Case studies of IoT implementation in aquaculture

The combination of Internet of Things (IoT) in aquaculture has discovered significant opportunity in enhancing harvest, aiding sustainability and idealizing resource utilization. Numerous circumstance studies from several areas and aquaculture backgrounds show the real -world applications and uses of IoT techniques in the industry. One distended case study is the implementation of IoT techniques in shrimp farming in Taiwan. The project elaborated the aligning of IoT sensors to perceive real-time water quality levels, which comprises parameters such as salinity, pH, temperature, and dissolved oxygen levels. These parameters are vital in upholding an ideal condition for shrimp growth. The IoT system practices a cluster of wireless sensor networks (WSNs) and cloud storage data analytics to bid aqua farmers with hands-on revelations and premature indications about water quality weakening. This empowered farmers to take remedial steps, such as varying ventilation or water flow, to stopover fatalities due to deficiency of good water conditions. The consequence was an obvious expansion in shrimp survival rates and yield efficiency, along with a decrease in energy prices related with water treatment (Chen et al., 2020). One more illustrious example is from Norway, where IoT technology has been abundantly merged in salmon farming. In this case, IoT sensors were linked in the fish pens for nonstop nursing of health of the fish and the environmental circumstances of the water. The sensors presented instant data on water quality, fish behavior and even fish activities with the pens. Examining this data, farmers can recognize early signals of diseases such as sea lice attacks or irregular behavior, which postulates stress. This permitted for more meticulous and appropriate interferences, which abridged the spread of diseases and diminished the requirement for antibiotics. Moreover, the poised data was shared into a centralized structure that allows prognostic examination, thus increasing decision-making and resource management (Petersen et al., 2020). Besides fish and shrimp farming, IoT has also been laboring in the guideline of water quality in freshwater aquaculture systems. In a case study from India, IoT sensors were active in freshwater ponds to perceive several water quality limits, including ammonia, nitrite and nitrate levels, which are decisive for fish well-being. The system was built to instinctively induce signs if there is any exceptional of parameters above the ideal limits, prompting farmers to yield counteractive solutions, such as regulating feed levels or energizing aeration systems. The real-time observatory system also allowed farmers to track and accomplish their pond situations, declining the time and labor required for physical checks. The result boosted fish development rates, lessened mortality and advances efficacy in source utilization (Srinivas et al., 2019). A further advanced example is invented in the extension of an IoT-dependent smart aquaculture system in Singapore. This system syndicates plentiful IoT technologies, which comprises sensor networks, machine learning and cloud computing, to provide a fully self-directed aquaculture farm. The system witnesses all crucial parameters such as water temperature, salt content, and dissolved oxygen, which certainly adapted environmental conditions to idealize fish well-being and development. In addition, the system merged AI-Powered analytics to predict the impending necessities of the farm, which comprises feed, water treatment, and energy utilization. This method not only enhances yield but also shrinkages the ecological footprint of aquaculture activities, contributing to sustainable practices in the industry (Tan et al., 2022). These case studies highlight the reformatory effect of IoT techniques on the aquaculture industry. Rapid data, predictive analytics and mechanization lets aquaculture measures to be extra productive, sustainable and tough. The competence to spot ecological conditions and fish well-being in present not only boosts yield but also decreases risks allied with disease eruptions and ecological degradation. However, encounters such as large early venture costs, complication in integration, and data security concerns remain, demanding keen observation and implementation of IoT solutions for various aquaculture frameworks (Dahl & Bäckström, 2023).

Challenges and limitations in IoT deployment

The fusion of Internet of Things (IoT) into aquaculture diplomacies has the ability to convert the industry by refining observation, enhancing working proficiencies, and certifying the well-being and productivity of aquatic species. Though, the placement of IoT in aquaculture grants its own set of difficulties and limitations that should be measured for prolific installation. One of the main problems is the higher charge of introductory disposition, which contains the positioning of sensors, data assessing units and communication structure. These expenditures steep for minor and medium sized aquaculture businesses, making it hard to gauge IoT resolutions across the industry (Kumar et al., 2024). Besides, energy feasting is the other important alarm. IoT devices, chiefly in rural and offshore aquaculture farms requests a stable power distribution. The reliance on renewable energy assets such as solar and wind makes modifications in power obtainability, which complicates the nonstop process of IoT systems (Mishra et al., 2021). Also, data provisions and grouping signifies substantial complications in IoT measures. The complete size of data formed by IoT sensors, grasps a variation of parameters from water quality to fish health biomarkers, challenging well sustained storage, assessment and inspection (Pal et al., 2024). The problem of directing faultless high occurrence data often imposes strong computational structure, which may not be readily accessible in all aquaculture contexts. This rises apprehensions about data safety and secrecy, as IoT systems are susceptible to cyberbullying that could distract events leading to illegal admittance to faint data (Zhou et al., 2022).

Also, the constrained linking in rural aquaculture areas characterizes other important limitation. Many aquaculture farms, especially in seashore or remote areas find hard with lack of internet connectivity, that interrupts the instant transfer of data from IoT devices to central control systems (Nguyen et al., 2021). This issue is worsened by the severe ecological situations in those areas, that impact the durability and participation of IoT hardware, which includes, erosion, water depletion and dangerous temperatures

(Rana et al., 2020). Besides technical experts and human resources causes obstacles in placement of IoT in aquaculture. Many aquaculture farmers and workers may lack the technical skills (Sharma et al., 2022). As the techniques changes quickly, current training and provision are required to make sure that utilizers efficiently implementing IoT into their prevailing operations. To conclude, the calibration and operational efficiency of IoT devices along various sellers and stages present a blockade to immediate incorporation. The absence of shared values delays data transferring between devices and systems, warning the scalability of IoT resolution across the aquaculture field (Pereira et al., 2021).

Sustainable and smart aquaculture practices

Sustainable aquaculture methods are important for warranting the long-term feasibility of the industry while reducing negative ecological effects. With the rapid progression of technology, the incorporation of the Internet of Things (IoT) with aquaculture has arisen as a turning point in endorsing sustainable and inventive methods. IoT techniques, which include sensors, self-governed systems and instant observatory tools, allows aquaculture activities to idealize resource management, improve yield efficacy and reduce ecological risks. These inventive techniques aid the collection of huge-scale instant data from fish farms, that is used to observe water quality, feed efficacy, fish well-being and ecological circumstances. Among the main obstacles faced by traditional aquaculture, regulating water quality is crucial for the health and development of aquatic species. IoT enabled sensors uninterruptedly quantify parameters such as temperature, pH, dissolved oxygen, ammonia, and salt levels in real time (Yoo & Jeong, 2020). This constant observation make sure that any alterations from ideal conditions are sharply identified and rectified decreasing the alarm of disease outbursts and diminishing the required for chemical involvements, such as antibiotics or pesticides, which damages both the ecosystem and human health.

As well, IoT systems assists in detecting initial warnings of diseases or stress in fish, permitting farmers to take defensive measures rather than responsive treatments, which are expensive and ecologically damaging (Zhou et al., 2022). Another important benefit of IoT incorporation in aquaculture is the idealization of feed regulation. Overfeeding is a general issue in fish farming, which leads to water pollution and unproductive source usage. Organizing IoT-supported feeding systems that are connected to instant information on fish behavior and water conditions allows aqua culturist to modify feeding time intervals and amount of feed, to make sure ideal fish feeding and waste reduction, thus enhancing feed conversion ratios (Jin et al., 2021). These innovative feeding systems not only decrease functional charges but also reduced the ecological impact of surplus nutrients, leading to eutrophication and other environmental issues in aquatic environments. Besides, IoT techniques has a critical role in encouraging sustainable energy utilization in aquaculture. Augmenting energy ingestion for water distribution, aeration and lighting based on instant data and prognostic analytics allows farms to decrease carbon footprint and functional costs (Zhang et al., 2023).

Also, the use of renewable energy sources, such as solar panels incorporated with IoT systems, supplementarily improves the sustainability of fish farming practices, makes them stronger against alterations in energy charges and environmental obstacles. The wide application of IoT techniques also endorses data-driven decision-making allowing aquaculture experts to enhance farm administration through predictive examinations. The huge data provided by IoT sensors were examined with the help of machine learning algorithms to identify trends, idealize functions and make prior decisions on species choice, stocking density and water quality management (Rodrigues et al., 2021). This transfer near data-aided methods parallels with the aim of precision aquaculture, that increases efficiency simultaneously decreasing environmental deprivation and resource intake.

Conclusion

IoT technologies, which include sensors, instant monitoring systems, and mechanized controlling pathways, can idealize ecological conditions, improve resource management and warrant the health and well-being of aquatic species. The use of IoT in nursing water quality, feeding patterns and disease control empowers more precise decision-making, waste reduction and improving the cost-effectiveness of aquaculture farms. Besides, the instant data obtained information supports the growth of predictive models, promoting proactive management strategies. Though, to fully comprehend these uses, obstacles such as construction expenses, data secrecy concerns, and the need for expertise must be considered. In conclusion, the fruitful incorporation of IoT in aquaculture will need association among industry shareholder, uninterrupted technological improvements and real-world policy outlines to encourage sustainable approaches and certify long-term industry development.

References

Abdulkareem, K. H., et al. (2022). IoT-based smart aquaculture: A review of applications and challenges. *Aquaculture, 546*, Article 737416.

Ali, M. A., Rao, P. K., & Li, Z. S. (2020). Cellular connectivity for aquaculture IoT systems: Challenges and opportunities. *Journal of Aquaculture Engineering, 12*(3), 341–356.

Chen, J., Sun, R., Pan, C., Sun, Y., Mai, B., & Li, Q. X. (2020). Antibiotics and food safety in aquaculture. *Journal of Agricultural and Food Chemistry, 68* (43), 11908-19.

Dahl, D., & Bäckström, Å. I. (2023). Meeting, moving, mastering - A text analysis of the aesthetic attractions of 'Wild Swimming'. *International Journal of Aquatic Research and Education, 14*(1), 12.

Duan, X., et al. (2020). Edge computing for aquaculture: An intelligent framework for smart fish farming. *Journal of Aquaculture Research & Development, 11*(5), 655–667.

FAO. (2020). The state of World fisheries and aquaculture 2020: Sustainability in action. *Food and Agriculture Organization of the United Nations.*1-244.

Ganguly, S., Sankar, P. M. G., & Chen, R. A. J. (2020). Securing IoT data in aquaculture environments. *Aquatic Science and Technology Journal, 25*(2), 55–67.

Gubbi, J., et al. (2013). Internet of Things (IoT): A vision, architectural elements, and future directions. *Future Generation Computer Systems, 29*(7), 1645–1660.

Hoch, E., Sidi, Y., Ackerman, R., Hoogerheide, V., & Scheiter, K. (2023). Comparing mental effort, difficulty, and confidence appraisals in problem-solving: A metacognitive perspective. *Educational Psychology Review, 35*(2), 61.

Hossain, M. S., Hossain, M. S. M. A., & Wu, M. R. (2018). Efficient communication protocols for IoT in aquaculture. *International Journal of Wireless Communications, 9*(4), 124–139.

Jiang, Y., Lee, M. R. M., & Zhao, C. R. P. (2021). Low power IoT networking for sustainable aquaculture. *Journal of Agricultural and Food Technology, 43*(8), 1152–1164.

Jin, W., Lai, Y., & Zhu, F. (2021). Effect of dietary fucoidan on innate immune response of Procambarus clarkii and disease resistance against white spot syndrome virus. *Aquaculture, 534*, 736233.

Kumar, J., Gupta, R., Sharma, S., Chakrabarti, T., Chakrabarti, P., & Margala, M. (2024). Iot-enabled advanced water quality monitoring system for pond management and environmental conservation. *IEEE Access, 12*, 58156-67.

Lee, C. Y., et al. (2020). Real-time monitoring and management system for aquaculture using internet of things. *Journal of Aquaculture Research & Development, 11*(2), 123–135.

Li, X., et al. (2022). IoT-based smart aquaculture systems: Review and future directions. *Aquaculture Engineering, 98*, 1–11.

Mishra, B. K., Kumar, P., Saraswat, C., Chakraborty, S., & Gautam, A. (2021). Water security in a changing environment: Concept, challenges and solutions. *Water, 13*(4), 490.

Muthukumar, V., Sivakami, R., Venkatesan, V. K., Balajee, J., Mahesh, T. R., Mohan, E., & Swapna, B. (2023). Optimizing heterogeneity in IoT infra using federated learning and blockchain-based security strategies. *International Journal of Computers Communications & Control, 18*(6).

Nguyen, H. Y., Trinh, T. L., Baruah, K., Lundh, T., & Kiessling, A. (2021). Growth and feed utilisation of Nile tilapia (*Oreochromis niloticus*) fed different protein levels in a clear-water or biofloc-RAS system. *Aquaculture, 536*, 736404.

Oliveira, R. F., Pereira, C. J., & Oliveira, H. D. (2019). Wireless sensor networks for aquaculture IoT. *IoT and Wireless Communication Networks, 11*(2), 129–143.

Pal, D., Jana, S., & Roy. (2024). Water Quality Prediction in Aquaculture (WQPA) Using Machine Learning and Internet of Things. In *International Conference on Smart Systems and Wireless Communication* (pp. 589–598). Singapore: Springer Nature Singapore.

Patel, A., & Kumar, P. (2021). IoT applications in aquaculture: A review. *Aquaculture Research, 52*(8), 3296–3307.

Pereira, I. M., de Senna Carneiro, T. G., & Figueiredo, E. (2021). Understanding the context of IoT software systems in DevOps. In *2021 IEEE/ACM 3rd International Workshop on Software Engineering Research and Practices for the IoT (SERP4IoT)* (pp. 13–20). IEEE.

Petersen, J. K., Taylor, D., Bergström, P., Buer, A. L., Darecki, M., Filippelli, R., Gren, M., Hasler, B., Holbach, A. M., Nielsen, P., & Petersen, L. K. (2020). *Policy guidelines for implementation of mussel cultivation as a mitigation measure for coastal eutrophication in the Western Baltic Sea.* Danish Shellfish Centre, National Institute of Aquatic Resources.

Rana, K. M., Ahammad, K., & Salam, M. A. (2020). Bioinformatics: scope and challenges in aquaculture research of Bangladesh-a review. *International Journal of Agricultural Research, Innovation and Technology, 10*(2), 137-45.

Raza, U., Farhan, M. S. J., & Mehmood, S. W. K. C. S. (2017). Data analytics in aquaculture IoT using edge computing. *Aquaculture Engineering Journal, 19* (6), 201–211.

Rodrigues, J. F, Jr., Florea, L., De Oliveira, M. C., Diamond, D., & Oliveira, O. N., Jr. (2021). Big data and machine learning for materials science. *Discover Materials, 1*(1), 12.

Sharma, A., Krishnani, K. K., Prasad, K. P., Verma, A. K., Kunjir, S. N., Rathod, R. H., ... Soni, S. (2022). Percolation Tank Based Aquaculture: Entrepreneurship Opportunity for Rural Tribal Youth. *Sushrirekha Das and Menaga. Meenakshisundaram*, 15.

Siti, A. R., Rafie, B. M. Z., & Ahmed, S. R. H. (2020). IoT systems in aquaculture: Challenges and future directions. *Journal of Aquatic Science, 36*(1), 32–41.

Srinivas, D., Venkatrayulu, C., Swapna, B., Swathi, A. V., & Venkateswarlu, V. (2019). Studies on socio-economic profile, problems and constraints of shrimp farmers in nellore district of Andhra Pradesh, India. *Asian Journal of Science and Technology, 10*(06), 9731-5.

Tan, K., Zhang, H., Li, S., Ma, H., & Zheng, H. (2022). Lipid nutritional quality of marine and freshwater bivalves and their aquaculture potential. *Critical Reviews in Food Science and Nutrition, 62*(25), 6990–7014.

Tao, Z., et al. (2019). Biometric sensors for fish health monitoring in aquaculture. In *Fish science and technology.*

Tao, F., et al. (2020). Digital twin and cyber-physical systems in the intelligent aquaculture. *Computers in Industry, 123*, Article 103302.

Vijayan, R., et al. (2020). Challenges in IoT adoption in small-scale aquaculture farming. *International Journal of Aquaculture and Fisheries, 9*(3), 76–84.
Yoo, S. L., & Jeong, J. C. (2020). Safe navigation distance between marine routes and aquaculture farms in South Korea using Gaussian mixture model. *Sensors, 20*(5), 1246.
Yuan, W., Christie-Oleza, J. A., Xu, E. G., Li, J., Zhang, H., Wang, W., Lin, L., Zhang, W., & Yang, Y. (2022). Environmental fate of microplastics in the world's third-largest river: basin-wide investigation and microplastic community analysis. *Water Research, 210,* 118002.
Zhang, Y., et al. (2021). Real-time water quality monitoring system for aquaculture based on IoT. *Sensors, 21*(14), 4767.
Zhang, M., et al. (2022). Artificial Intelligence and IoT integration in sustainable aquaculture. *AI in Aquaculture, 1*(1), 45–58.
Zhang, Z., Cui, Y., & Zhang, Z. (2023). Unequal age-based household carbon footprint in China. *Climate Policy, 23*(5), 577-92.
Zhou, D., et al. (2021). Internet of things for sustainable aquaculture: An overview of applications. *Sensors, 21*(7), 2405.
Zhou, X., Fang, J., Shi, S., Yang, L., Xie, B., & Huang, L. (2022). Ecosystem services of ecosystem approach to mariculture: Providing an unprecedented opportunity for the reform of China's sustainable aquaculture. *Frontiers in Marine Science, 9,* 909231.

10

OMICS technologies for aquaculture

Introduction to OMICS and aquaculture

The 21st century has perceived a standard change in biological research, determined by the conjunction of highly productive techniques and computational biology, known as OMICS. OMICS includes a group of interlinked fields, which includes genomics, transcriptomics, proteomics, metabolomics, and others that aid all-inclusive visions into the molecular pathways that regulate life processes (Fig. 10.1). In regard to aquaculture, a fast-growing food production sector that offers about half of the worldwide fish production, the incorporation of OMICS has progressively become crucial for enhancing yield, sustainability, and disease resistance demand. As aquaculture deepens to encounter the rising global claim for seafood, obstacles such as disease outbreaks, environmental pollutants, and the requirement for genetic modification need novel and combined biotechnological methods. OMICS techniques, in combination with artificial intelligence (AI), showcase an unparalleled occasion to unleash the multifaceted biological web within aquatic lives and their surroundings, eventually improving aquaculture practices from broodstock selection to harvest. Genomics, the initial pillar of OMICS, emphasizes the structure, function, evolution, and mapping of genomes. The use of next-generation sequencing (NGS) techniques has enabled the whole or comprehensive sequencing of various aquaculture species such as Atlantic salmon (*Salmo salar*), Nile tilapia (*Oreochromis niloticus*), and Pacific white shrimp (*Litopenaeus vannamei*). Genome-wide association studies (GWAS), quantitative trait locus (QTL) mapping, and selective breeding programs supported by genomic data are allowing accurate breeding for traits including growth rate, disease resistance, and ecological tolerance. For example, genomic selection in Atlantic salmon has suggestively decreased the frequency of transmissible pancreatic necrosis by recognizing resilient genotypes (Houston et al., 2014). Genomic datasets are also

Aquaculture with Artificial Intelligence. https://doi.org/10.1016/B978-0-443-33704-8.00006-9

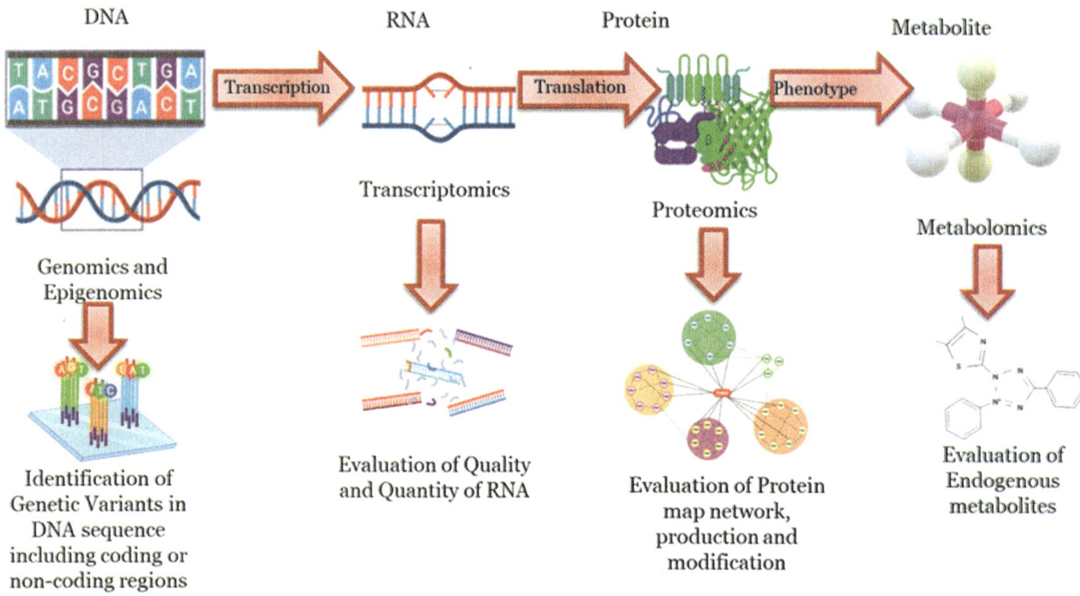

Figure 10.1 Molecular mechanism of omics technology used in aquaculture.

contributory in preserving genetic diversity, reducing inbreeding depression, and improving the flexibility of aquatic ecosystems to altering atmospheres. Transcriptomics covers the genomic landscape by seizing gene expression profiles underneath diverse physiological or ecological conditions. With the help of techniques such as RNA sequencing, transcriptomics researchers can recognize capable genes, governing webs, and mechanisms tangled in stress responses, pathogen defense, reproduction, and development; for instance, proportional transcriptomic analyses have clarified immune-related genes in shrimp unprotected from white spot syndrome virus (WSSV), providing goals for vaccine discovery and immunostimulant construction (Li & Xiang, 2020). Furthermore, time-series transcriptomic studies support the dynamic observation of responses to dietary interferences or salinity alterations, thus idealizing feed formulations and culture methods. Proteomics, the large-scale work of proteins, conduits the slit between genotype and phenotype by offering perceptions into active molecules arranging cellular physiology. Since proteins are the initial effectors of biological processes, their richness, posttranslational alterations, and connections are crucial for knowing the biology of culture organism. Proteomic approaches have discovered biomarkers for disease diagnostics, metabolic well-being, and multiplicative position in fish and crustaceans. For example, various protein expressions in serum and gill tissues have been related to bacterial and parasitic infections in catfish and carp (Rodrigues et al., 2021). Mass spectrometry-dependent proteomics, often combined with bioinformatics and systems biology, provides practical goals for functional genomics and therapeutic interventions. Metabolomics provides a good opinion on biological procedures by analyzing the tiny-molecule metabolites found inside cells, tissues, or biofluids. This OMICS layer showcases the organism's metabolic phenotype, offering instantaneous photographs of physiological position in response to diet, disease, or ecological fluctuations. Nuclear magnetic resonance (NMR) spectroscopy and liquid chromatography-mass spectrometry (LC-MS) are among the technologies laboring to characterize metabolomic outlines. In aquaculture, metabolic data have shown that the preparation of functional feeds augmented with probiotics, prebiotics, and herbal extracts enhances immunity and development rate. In addition, metabolomics supports translating the composite host–microbe connections in the gut environment that is important for larval development and feed adaptation efficiency (Wishart, 2019). Consolidative OMICS, or multi-OMICS methods, are specifically powerful in aquaculture, where various traits are the rule. Uniting data from genomics, transcriptomics, proteomics, and metabolomics, researchers construct wide-ranging webs that outline gene–protein–metabolite connections; thus, it leads to a system-level understanding. For instance, against the thermal stress, the amalgamation of transcriptomic and metabolomic datasets in tilapia has recognized monitoring circuits that alleviate oxidative damage and maintain cellular homeostasis (Y. Wang et al., 2019). Such consolidative tactics are also treasured in host–pathogen studies, as they unleash harmonious immune pathways and novel antimicrobial targets. The cooperation between OMICS and AI is transforming aquaculture by altering huge amounts of biological data into prognostic models and decision-making tools. Machine learning (ML) algorithms excavate high-dimensional OMICS datasets to recognize patterns, distinguish disease phenotypes, and forecast breeding values with improved accuracy. Deep learning (DL) outlines are being used in image-based phenotyping, ecological observation, and behavior tracking, with additional sustenance from molecular perspectives obtained from OMICS. AI- supported OMICS enables present, data-based aquaculture administration systems that are adaptable and robust to global obstacles, such as climate change, emerging diseases, and food insecurity.

Genomics: decoding the blueprint of aquatic species

Genomics, the wide-ranging study of an animal's complete genome and its relations, has transformed our knowledge of aquatic species by offering a vast blueprint of their biology. In aquaculture, genomics aids researchers and aquaculture practitioners to decode the genetic structure of key characteristics, which include growth rate, disease susceptibility, ecological tolerance, and breeding efficacy. Integrating high-end sequencing techniques, scientists are positioning the genomes of finfish, crustaceans, mollusks, and other frugally vital aquatic species with unparalleled accuracy. This genomic vision empowers the recognition of genetic markers linked with required phenotypes, thus helping marker-assisted selection (MAS) and genomic selection (GS) programs that rush breeding cycles and improve productivity in aquaculture systems (Houston et al., 2020). The usage of whole-genome sequencing (WGS) and transcriptomics has not only unleashed the architecture and activities of thousands of genes but also lean light on how gene expression is managed under different ecological and pathogenic pollutants. For example., genomic analysis of disease-resistant strains of shrimp and tilapia has led to the recognition of immune-related genes and single-nucleotide polymorphisms (SNPs) that have been bound for selective breeding (Yanez et al., 2014). Moreover, comparative genomics of species and populations offers serious understandings into evolutionary adaptation and speciation, providing direction for the maintenance of genetic diversity in farmed and wild stocks. The mixing of functional genomics, which includes CRISPR/Cas9-mediated gene editing, supplementarily covers these potentials by aiding the direct alteration of genes to improve specific characteristics. Though ethical and supervisory regulations are still transforming around their utilization in aquaculture. Genomic databases and bioinformatic tools now play an important role in handling and analyzing the huge amount of data obtained by sequencing projects. Inventiveness such as the Fish-T1K Project and Genome 10K, aims to list genomic data across an extensive range of aquatic species, developing cooperative research and comparative examinations (Sun et al., 2016). Combining genomics with AI is rising as a transformative method in aquaculture, where ML algorithms detect breeding outcomes, idealize feed formulations based on genotype, and observe population well-being via genomic biomarkers. As sequencing costs continue to decrease, the convenience and usefulness of genomics in aquaculture will extend, empowering accurate breeding, sustainable yield, and resistance against rising obstacles such as climate change and disease outbursts.

Transcriptomics: Understanding gene expression in aquatic organisms

Transcriptomics, the extensive field of RNA transcripts formed by the genome under the precise conditions or in a specific cell, acts an important role in learning the functional genomics of aquatic organisms (Fig. 10.2). This discipline allows researchers to seize the dynamic property of gene expression, which offers visions into how environmental aspects, disease, growth status, and nutritional levels manipulate the

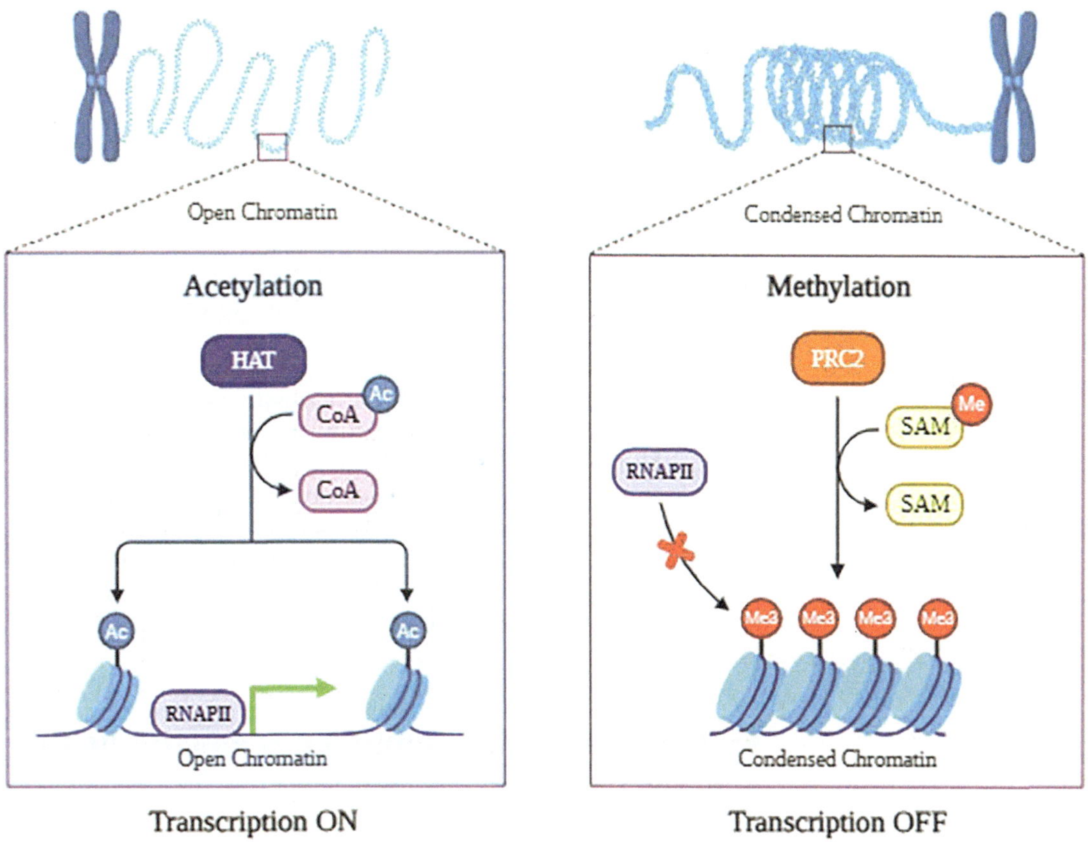

Figure 10.2 Mechanism of gene expression in aquatic organisms.

transcriptome. In the framework of aquaculture, transcriptomic learnings have become a necessary equipment for enhancing fish well-being, improving yield, and warranting environmental sustainability. The transcriptome of an aquatic organism showcases its physiological condition and adaptive responses to both biotic and abiotic stressors. Through RNA sequencing (RNA- Seq), the quantification of gene expression from tissues, time-points, and even in nonmodel organisms that have limited genomic expression is also possible. For instance, transcriptomic profiling in species such as Atlantic salmon (*S. salar*), zebra fish (*Danio rerio*), and tilapia (*Oreochromis* spp) has revealed important controlling genes associated with immune responses to pathogens such as *Vibrio* spp. *Streptococcus iniae* and *Aeromonas hydrophila* (Palti et al., 2015; T. Wang et al., 2019). Likewise, studies have identified heat shock proteins, cytokines, and antimicrobial peptides that are uniquely expressed against thermal stress, hypoxia, and salinity changes offering capable biomarkers for ecological flexibility (Li et al., 2021).

Moreover, transcriptomic analyses are involved in nutritional genomics by explicating the molecular effects of dietetic interferences. This method allows the assessment of novel feed ingredients such as probiotics, prebiotics, and purposeful additives and their effect on metabolic pathways, gut well-being, and immune alteration (Calduch-Giner et al., 2012). In broodstock administration and larval growth, transcriptomics also aids in understanding the molecular signals elaborated in gametogenesis, embryogenesis, and organogenesis, thus optimizing hatchery practices (Yanez et al., 2014). The incorporation of transcriptomic data with other OMICS layers, namely genomics, proteomics, and metabolomics, supplements the prognostic control of structural biology representations in aquaculture. When integrated with ML and AI, these datasets enable the growth of accurate aquaculture methods that include real-time disease diagnostics and adaptive feeding commands. As sequencing becomes cost-effective and bioinformatics tools become more available, transcriptomics is established to play a major role in customized aquaculture, selective breeding programs, and ecological observation.

Proteomics: Protein profiling in aquatic health and development

Proteomics, the significant study of proteins, acts as an essential tool in illuminating the health and development of aquatic organisms, especially within the enlarging area of aquaculture. As proteins are the prime factors of cellular function, knowing their expression, alterations, and connections provides more visions on physiological and pathological progressions in aquatic species (Fig. 10.3). Protein sketching permits researchers to detect biomarkers suggestive of stress, disease, nutrition, reproductive status, and ecological acceptance, enabling it to be an influential analytical and predictive tool. Current developments in MS-dependent proteomics, such as tandem MS (MS/MS) and label-free quantification, have allowed the high-end detection and quantification of proteins in complex biological matrices such as fish serum, gill tissue, liver and muscle (Zhang et al., 2020). In aquaculture, these methods have been helpful in characterizing the host immune response to pathogens, thus helping the establishment of vaccines and immunostimulants (Zhao et al., 2021).

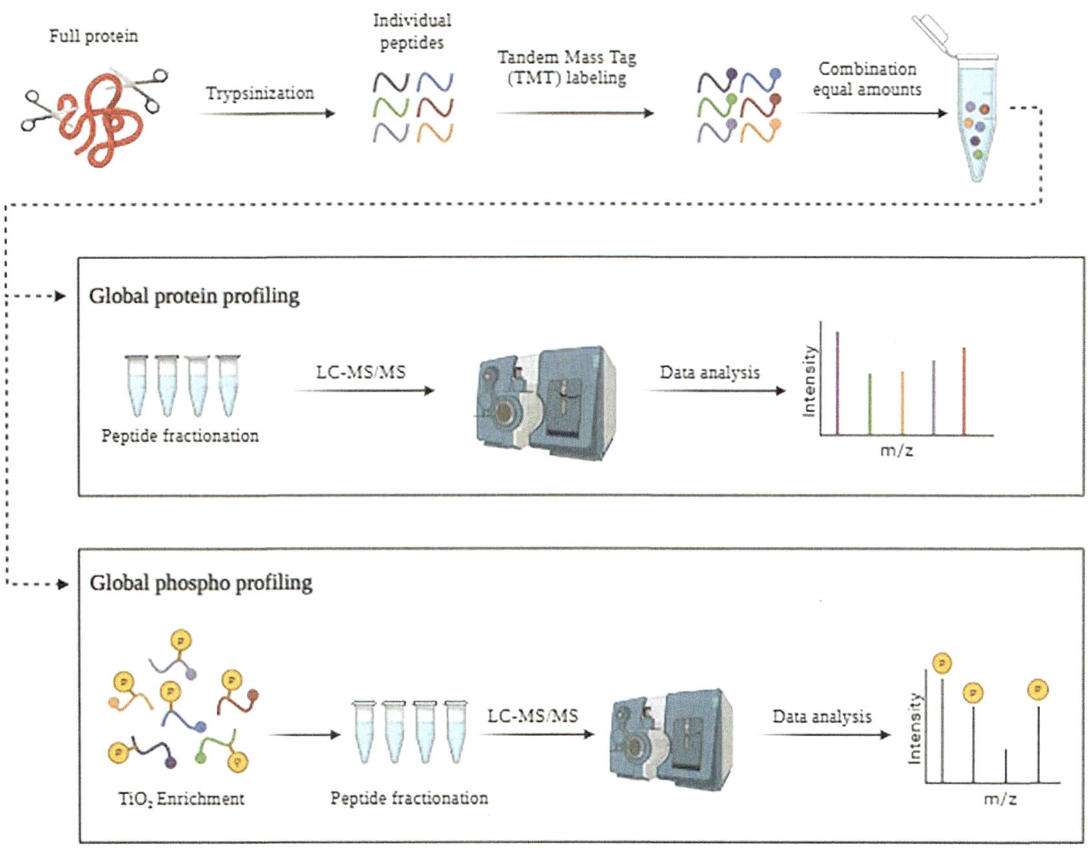

Figure 10.3 Mechanism of protein profiling in aquatic organism.

In addition, proteomics has poured attention to mechanisms behind the growth maintenance, metamorphosis, osmoregulation, and tissue regeneration in prime aquaculture species, which include tilapia, shrimp, and salmon. The amalgamation of proteomics with other omics platforms, such as transcriptomics and metabolomics, additionally improves our knowledge of gene–protein metabolite webs in aquatic systems, especially in stress conditions such as temperature variations, hypoxia, salinity fluxes, and pollution (Li et al., 2019). Furthermore, AI-assisted bioinformatics tools are now utilized to decode compound proteomic datasets, simplifying pattern identification and prognostic modeling for health observation and accurate breeding. The blend of proteomic technologies with AI holds scope for early disease detection, nutritional idealization, and choosing strong genotypes, eventually directing sustainability in aquaculture. As aquaculture changes into a more data-based and technically progressive sector, proteomics will continue as a center in making sure the fish health, enhancing yield, and reducing losses linked with disease outbursts (Kumar & Shanker, 2022).

Metabolomics: Insights into metabolic pathways and nutritional physiology

Metabolomics, a fast-growing field among the omics sciences, plays an important role in explaining metabolic pathways and considers nutritional physiology in aquaculture species (Fig. 10.4). It encompasses the wide-ranging study of metabolites, small molecular intercedes and outcomes of metabolism within a biological system under precise conditions. Despite genomics or transcriptomics, which offer still polaroids of capabilities and gene expression, metabolomics provides a lively opinion of a creature's physiological state, showing real-time responses to ecological, nutritional, or pathological stimuli (Wishart, 2019). In aquaculture, metabolomics empowers the detection of important biochemical markers allied with development, stress tolerance, disease resistance, and feed efficacy, thus simplifying the expansion of meticulous aquaculture approaches. In aquaculture, one of the biggest powers of metabolomics relies in its capacity to position metabolic pathways and unleash how nutrients are managed and applied. For example, by using smearing NMR spectroscopy or MS equipped platforms, researchers enumerate amino acids, lipids, and energy metabolites to measure the nutritional status and metabolic health of farmed fish or shrimp (Dettmer et al., 2007; Zhang et al., 2022). These data

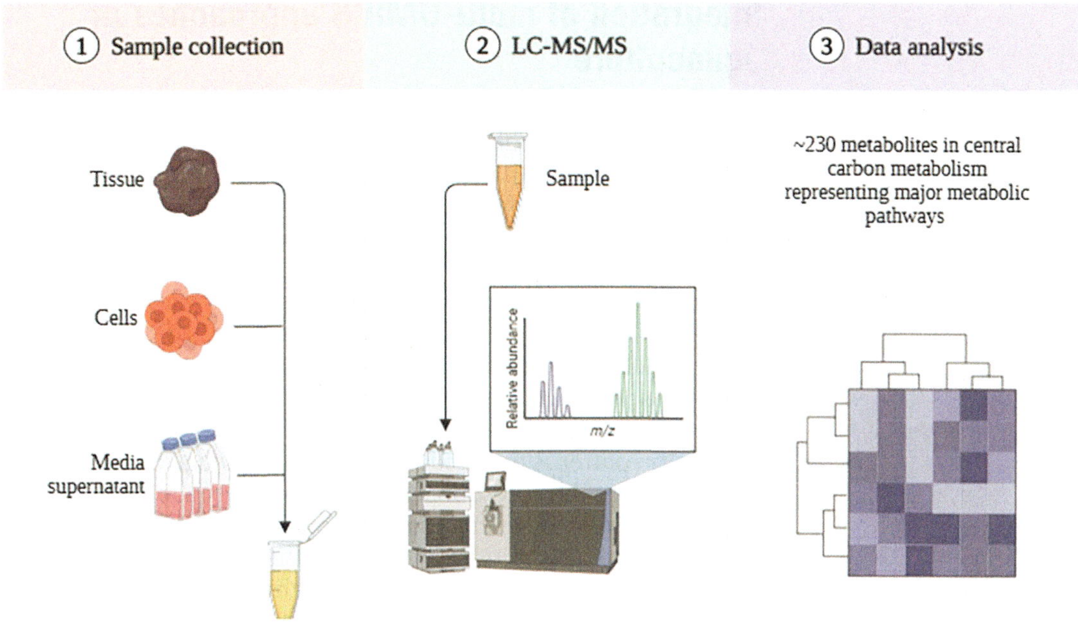

Figure 10.4 Basic metabolic pathways in aquatic organisms.

are combined with transcriptomic and proteomic evidence to build a systems biology pattern that notifies feed formulation and health administration tactics. In precise, metabolomic sketching has been vital in learning nutrient–gene interactions, metabolic variations to other protein sources, and the effect of probiotics and prebiotics on digestive health. Furthermore, metabolomics enables the premature uncovering of metabolic illness and oxidative stress, permitting for practical intercessions to improve survival and yield. As aquaculture grows in the direction of sustainability and efficiency, metabolomics offers a requisite equipment for illustrating the effects of ecological stressors, such as temperature alterations, salinity variations, and hypoxia on aquatic organisms. When combined with AI and ML algorithms, metabolomic datasets are extracted to determine prognostic biomarkers and augment culture conditions in the present (Johnson et al., 2017). This incorporation parallels well with the insights of a wise aquaculture frame that are adaptive, approachable, and resource-efficient. Arresting the composite biochemical fingerprints of aquatic organisms, metabolomics not only extends our knowledge of nutritional physiology but also quickens the growth of tough aquaculture methods for forthcoming food security.

Integration of multi-OMICS approaches in aquaculture

The incorporation of multi-OMICS methods in aquaculture is transfiguring our learnings of aquatic organisms and ecosystems, allowing exact, data-driven policies to enhance well-being, yield, and sustainability. Multi-OMICS incorporates a group of highly advanced techniques, which include genomics, transcriptomics, proteomics, metabolomics, and epigenomics that cooperatively offer a complete interpretation of biological systems (Hasan et al., 2021). In aquaculture, genomics has allowed the detection of genes linked with disease confrontation, development levels, and stress lenience, easing selective breeding programs (Yáñez et al., 2015). Transcriptomics leads this by seeing through gene expression patterns exposed to altering ecological or pathological conditions, such as responses to pathogen attack or ecological pollutants. Proteomics matches these discoveries by expounding the functional protein outputs and their exchanges, offering awareness into physiological mechanisms important to immune activities, reproduction, and metabolism (Rodrigues et al., 2017). Metabolomics deals with small molecule metabolites, augmenting another layer by apprehending the biochemical fingerprints of physiological situations and ecological revelations, contributing initial biomarkers for disease identification and nutritional shortages (Bobe & Goetz, 2001). The combination of these OMICS layers allows the growth of extrapolative models and customized interferences.

The integration of these OMICS layers enables the development of predictive models and tailored interventions, particularly when combined with AI tools. AI algorithms analyze complex, multidimensional OMICS datasets to identify patterns and correlations that would be otherwise undetectable through traditional statistical methods (Nguyen et al., 2024). For example, ML predicts host–pathogen interactions or identifies biomarkers for specific diseases, significantly improving biosecurity and health management in aquaculture. In addition, combined OMICS methods have been verified as influential in microbiome studies, where metagenomics and metatranscriptomics help discover the parts of gut and ecological microbiota in nutrient acclimatization, immunity, and development in fish and shellfish (Liu et al., 2022). Inducing AI to synthesize data across OMICS platforms permits researchers and industry consultants to make informed choices on feed formulation, breeding, disease control, and environmental observation. Finally, the junction of multi-

OMICS techniques with AI is navigating aquaculture into a novel era of accuracy in aquaculture, considering sustainability, improved animal well-being, and enhanced economic feasibility. Upcoming research should aim on data calibration, consolidative bioinformatics gears, and the growth of user-accessible fields to make sure that the uses of OMICS-based aquaculture reach both researchers and practitioners in the field.

Artificial intelligence in OMICS data analysis

AI has arisen as a reformative tool in the examination of OMICS data genomics, transcriptomics, proteomics, and metabolomics, especially in regard to aquaculture, where the production of high-end data has overtaken outdated investigative approaches. The collaboration of AI techniques such as ML, DL, and natural language processing (NLP) has allowed the invention of composite outlines, prophetic modeling, and instantaneous decision-making that were formerly unachievable. In aquaculture, genomics data empowered by AI has enabled the detection of disease-resistant genotypes, allowing the growth of discriminating breeding programs for strong fish strains. For example, ML algorithms such as arbitrary forest and support vector machines have been used to categorize gene expression profiles associated with pathogen response in shrimp and fish, improving disease monitoring and mitigation approaches. Transcriptomics, which studies the mRNA expression, uses AI by gathering coexpressed genes and envisaging regulatory grids, thereby providing visions into stress physiology and ecological adaptation (Díaz Domínguez, 2021). Proteomics and metabolomics, characterized by huge amounts of information and composite molecular interactions, have seen notable developments through the utilization of deep neural networks and unsupervised learning, facilitating the detection of biomarkers for nutritional idealization and water quality observation (Trivedi et al., 2022).

In addition, the use of AI in consolidative multi-OMICS examination has enabled the wide-range learning of biological systems in aquaculture species, helping systems biology methods for sustainable production. Equipment such as AutoML and AI-based bioinformatics platforms, such as DeepOMICS and DeepSEA, have efficient data preexamining, feature choice, and functional explanation, reduced human blunder, and augmented theory generation (Zou et al., 2019). In addition, AI facilitates the incorporation of instantaneous data from mobile biosensors and ecological observation systems, thus linking OMICS insights with aquaculture management choices, such as feeding time tables and health interferences. As aquaculture enlarges to meet global food demands, the interaction between OMICS techniques and AI holds proven results for accurate aquaculture, developing innovations in breeding, nutrition, and environmental sustainability. Continued progress in understandable AI will further enhance the interpretability and trustworthiness of calculations, enabling AI-integrated OMICS to reach a milestone of wise aquaculture systems.

Applications in disease diagnostics and health management

The claim of OMICS skills, genomics, transcriptomics, proteomics, and metabolomics in aquaculture has transfigured disease diagnostics and well-being supervision by allowing quick, accurate, and extrapolative scrutiny of host–pathogen connections, immune responses, and disease biomarkers. These gears deliver unparalleled awareness into the molecular foundations of disease outbursts and are vital in supervisory battered intercessions. Genomics, predominantly NGS, eases the detection of disease-resistant strains through GWAS, which has led to the expansion of choosy breeding agendas in species such as Atlantic salmon and tilapia. For instance, Houston et al. (2014) established the utility of GWAS in recognizing QTLs allied with confrontation to communicable pancreatic necrosis virus in salmon, flagging the way for MAS. Transcriptomics, using RNA-Seq technology, delivers a dynamic snap of gene expression fluctuations during contamination. Case studies, such as the one by Zhang et al. (2023), have voyaged the transcriptomic profiles of *L. vannamei* shrimp diseased with WSSV, see through the upregulation of immune-related genes and the identification of novel antiviral pathways, which aid as possible investigative biomarkers. Proteomics balances this by classifying and enumerating proteins involved in immune response pathways.

For instance, Li and Xiang (2020) employed mass spectrometry-based proteomics to investigate disease progression in groupers infected with nervous necrosis virus (NNV), resulting in the identification of differential protein markers that are valuable for early-stage diagnosis. Metabolomics provides insights into physiological alterations during disease states. In a recent case study, Kumar et al. (2021) used NMR spectroscopy to profile metabolites in rohu (*Labeo rohita*) infected with *A. hydrophila*. The study identified specific metabolic disruptions, including changes in amino acid and lipid profiles, which could be integrated into diagnostic tools for early detection. Integration of these OMICS platforms through systems biology and AI-based analytics enables real-time disease surveillance and health forecasting in aquaculture systems. For instance, AI models trained on multiomics data predict disease onset with high accuracy, supporting precision aquaculture practices. The fusion of OMICS with ML, as demonstrated in recent studies utilizing neural networks for pathogen detection from transcriptomic signatures, showcases the potential of this approach for noninvasive, high-throughput diagnostics (Moghadam et al., 2017).

OMICS in selective breeding and genetic improvement

OMICS machineries, including genomics, transcriptomics, proteomics, and metabolomics, are transfiguring the discriminatory breeding and genetic upgrading of aquatic species. These progressive molecular apparatus afford a thorough consideration of the genetic makeup, gene expression, and biochemical pathways that participate in growth, disease battle, and reproductive performance. The application of OMICS skills in aquaculture has suggestively improved the efficacy of breeding programs by permitting the assortment of greater traits with superior accuracy. Genomics, in precise, has played a key role in studying the genetic architecture of aquaculture species. For illustration, the sequencing of the genomes of species such as *S. salar* (Atlantic salmon) and *O. niloticus* (Nile tilapia) has provided valued indication about the genes related to growth rates, disease resistance, and environmental flexibility. This data permits the identification of genetic markers used for MAS, thereby quickening the upbringing process and confirming that necessary traits are approved on to the next generation more competently (Gonen et al., 2024).

Also, the application of transcriptomics has allowed the examination of gene expression profiles underneath numerous ecological circumstances or stressors, serving breeders to choose personalities that prosper in challenging aquaculture systems (F. Wang et al., 2019). A case study illustrating the fruitful application of OMICS in selective breeding is the enhancement of disease resistance in shrimp farming. WSSV is an important danger to shrimp aquaculture, ensuing in considerable losses. Using genomic and transcriptomic tools, researchers have recognized immune-related genes connected with the battle between WSSV and shrimp. These genetic markers have been combined into breeding programs, allowing the expansion of shrimp lines with boosted disease resistance, thus refining the sustainability and productivity of shrimp farming. This method highlights the latent of OMICS to not only improve frugally vital traits, such as progress, but also address critical encounters, including disease epidemics in aquaculture.

Besides, proteomics and metabolomics offer harmonizing insights by analyzing the proteins and metabolites involved in growth, metabolism, and stress responses. For example, proteomic analyses have aided in classifying proteins linked with muscle development in fish, conducive to the expansion of strategies for refining feed conversion competence and growth rates (Li et al., 2019). Metabolomics, on the other hand, affords perceptions into how several feed constituents or environmental settings affect the metabolic pathways of aquatic species, aiding the optimization of nourishing strategies and more effective aquaculture management (Wang et al., 2020). Inclusive of all, the integration of OMICS skills into selective breeding programs signifies a transformative method for genetic enhancement in aquaculture. Allowing an exact assortment of desirable traits and providing unfathomable understandings into the molecular reinforcements of aquaculture species, these skills offer unparalleled chances to augment output, sustainability, and disease battle in aquaculture systems.

Challenges and future directions

OMICS technologies, encircling genomics, transcriptomics, proteomics, metabolomics, and microbiomics, grasp great potential for proceeding the aquaculture industry. These skills provide all-inclusive visions into the biological and ecological features that disturb aquatic species, simplifying perfection in breeding, disease supervision, and sustainable performance. Though their request in aquaculture faces several encounters, chiefly in terms of data incorporation, cost, and the user-friendliness of cutting-edge knowledge.

One of the main challenges is the difficulty of data produced through OMICS skills. These datasets are frequently massive, varied, and high-dimensional, assembling it tough to extract practical visions without progressive computational tools (Wang & Srinivasan, 2017). While progressions in AI and ML are enhancing the capability to examine and understand OMICS data, there is still an important need for better algorithms and models custom-made to the unique stresses of aquaculture (Li et al., 2019). Furthermore, the combination of multi-OMICs data (e.g., genomics with proteomics and metabolomics) presents another layer of difficulty, necessitating cultured bioinformatics tools and interdisciplinary teamwork. Cost remains a large barrier to the extensive acceptance of OMICS machineries in aquaculture, particularly for small- and medium-scale farms in emerging countries. High-throughput sequencing technologies, progressive metabolomic assays, and other high-end instrumentation are excessively expensive, restraining their availability (Huang & Rust, 2021). Furthermore, particular proficiency in bioinformatics and data interpretation is regularly essential to make sense of the data, added increasing the cost and intricacy for aquaculture experts. Another challenge is the lack of consistent methodologies across diverse provinces and species. Aquaculture includes a variety of species, each with exclusive biological characteristics, and the approaches for sampling, sequencing, and data analysis vary largely. This lack of reliability delays cross-species evaluations and the growth of generalizable models for aquaculture management (Dunlap, 2020). Looking to the future, several guidelines promise to improve the impact of OMICS skills in aquaculture. First, the development of more affordable and portable technologies will democratize access to OMICS tools, chiefly for small-scale and developing-world aquaculture operations. Developments in low-cost sequencing platforms and shrunken diagnostic devices are previously on the skyline, and further inventions in this zone are likely to drive noteworthy perfections in aquaculture sustainability (Xu et al., 2021).

Besides, the addition of OMICS with other emergent technologies, such as AI, sensor networks, and blockchain, has the possibility to transfigure aquaculture management. AI-powered analytics could permit real-time decision-making based on dynamic ecological and biological data, while blockchain could ensure traceability and transparency in the supply chain, improving sustainability. Such collaborations could augment fish health nursing, decrease resource consumption, and progress overall productivity. Finally, collaborative worldwide exertions to normalize OMICS protocols, share data, and create global databases will be indispensable for proceeding the application of these technologies. Creativity meant at creating open-access mines of genomic and transcriptomic data for aquaculture species will empower researchers and practitioners to shape upon each other's work, thus quickening origination and enlightening the sector's competence (Meuwissen et al., 2020).

References

Bobe, J., & Goetz, F. W. (2001). Cysteine protease inhibitor is specifically expressed in pre-and early-vitellogenic oocytes from the brook trout periovulatory ovary. *Molecular Reproduction and Development: Incorporating Gamete Research, 60*(3), 312–318.

Calduch-Giner, J. A., et al. (2012). Transcriptional assessment of the effect of dietary cholesterol and docosahexaenoic acid on the intestinal and hepatic transcriptomes of European sea bass (*Dicentrarchus labrax*). *British Journal of Nutrition, 108*(12), 2115–2128.

Dettmer, K., Aronov, P. A., & Hammock, B. D. (2007). Mass spectrometry-based metabolomics. *Mass Spectrometry Reviews, 26*(1), 51–78.

Díaz Domínguez, D. A. (2021). Data structures and algorithms for analyzing DNA sequences in compressed space (Doctoral dissertation). Universidad de Chile.

Dunlap, M. (2020). Artificial Intelligence and the threat to free will. *Ex Nihilo, 2020*, 46.

Gonen, S., Benfey, T. J., & Garber, A. F. (2024). The genomic architecture of high temperature tolerance in a year class of Atlantic salmon. *Aquaculture, 578*, 740020.

Hasan, J., Lima, R. A., & Shaha, D. C. (2021). Fisheries resources of Bangladesh: A review. *International Journal of Fisheries and Aquatic Studies, 9*(4), 131–138.

Houston, R. D., Bean, T. P., Macqueen, D. J., Gundappa, M. K., & Jin, Y. H. (2020). Harnessing genomics to fast-track genetic improvement in aquaculture. *Nature Reviews Genetics, 21*(7), 389–409.

Houston, R. D., Taggart, J. B., Cezard, T., et al. (2014). Development and validation of a high density SNP genotyping array for Atlantic salmon. *BMC Genomics, 15*, 90.

Huang, M. H., & Rust, R. T. (2021). Engaged to a robot? The role of AI in service. *Journal of Service Research, 24*(1), 30–41.

Johnson, J. E., Bell, J. D., Allain, V., Hanich, Q., Lehodey, P., Moore, B. R., Nicol, S., Pickering, T., & Senina, I. (2017). The Pacific Island region: Fisheries, aquaculture and climate change. In , *Vol. 1. Climate change impacts on fisheries and aquaculture: A global analysis* (pp. 333–379).

Kumar, G., & Shanker, S. (2022). Artificial intelligence and omics technologies in sustainable aquaculture. *Frontiers in Marine Science, 9*, Article 842711.

Kumar, V., Roy, S., Behera, B. K., Bossier, P., & Das, B. K. (2021). Acute hepatopancreatic necrosis disease (AHPND): Virulence, pathogenesis and mitigation strategies in shrimp aquaculture. *Toxins, 13*(8), 524.

Li, C., et al. (2019). Integrated omics approaches in aquaculture research: a review. *Reviews in Aquaculture, 11*(4), 1101–1119.

Li, F., & Xiang, J. (2020). Recent advances in understanding the innate immune mechanisms of shrimp in response to viral infection. *Fish & Shellfish Immunology, 98*, 903–914.

Li, M., et al. (2021). Transcriptomic insights into thermal adaptation and stress response in marine fishes. *Frontiers in Marine Science, 8*, Article 651234.

Liu, J., Bienvenido, F., Yang, X., Zhao, Z., Feng, S., & Zhou, C. (2022). Nonintrusive and automatic quantitative analysis methods for fish behaviour in aquaculture. *Aquaculture Research, 53*(8), 2985–3000.

Meuwissen, M. P., Feindt, P. H., Midmore, P., Wauters, E., Finger, R., Appel, F., Spiegel, A., Mathijs, E., Termeer, K. J., Balmann, A., & de Mey, Y. (2020). The

struggle of farming systems in Europe: Looking for explanations through the lens of resilience. *EuroChoices, 19*(2), 4–11.
Moghadam, H. K., Johnsen, H., Robinson, N., Andersen, Ø., H. Jørgensen, E., Johnsen, H. K., Bæhr, V. J., & Tveiten, H. (2017). Impacts of early life stress on the methylome and transcriptome of Atlantic salmon. *Scientific Reports, 7*(1), 5023.
Nguyen, M. Q., Nguyen, D. M., Toan, T. T., & Dao, A. Q. (2024). Nanotechnology in aquaculture: Applications and challenges. *Journal of the Electrochemical Society, 171*(5). Article 057507.
Palti, Y., et al. (2015). Differential gene expression and SNP association with resistance to bacterial cold water disease in rainbow trout. *BMC Genomics, 16*(1), 1–14.
Rodrigues, J. G., Conides, A. J., Rivero Rodriguez, S., Raicevich, S., Pita, P., Kleisner, K. M., Pita, C., Lopes, P. F., Alonso Roldáni, V., Ramos, S. S., Klaoudatos, D., ... Villasante, S. (2017). Marine and coastal cultural ecosystem services: Knowledge gaps and research priorities. *One Ecosystem, 2*.
Rodrigues, P. M., Silva, T. S., & Esteves, E. (2021). Proteomics in aquaculture: applications and trends. *Aquaculture Reports, 20*, Article 100694.
Sun, Y., Zhang, Q., & Guo, W. (2016). Fish-T1K (transcriptomes of 1,000 fishes) project: Large-scale transcriptome data for fish evolution studies. *GigaScience, 5*(1), 18.
Trivedi, N., Erickson, H. E., Bala, V., Chhonker, Y. S., & Murry, D. J. (2022). A concise review of liquid chromatography-mass spectrometry-based quantification methods for short chain fatty acids as endogenous biomarkers. *International Journal of Molecular Sciences, 23*(21), 13486.
Wang, F., Li, S., Xiang, J., & Li, F. (2019). Transcriptome analysis reveals the activation of neuroendocrine-immune system in shrimp hemocytes at the early stage of WSSV infection. *BMC Genomics, 20*(1), 247.
Wang, T., et al. (2019). Transcriptomic profiling of the gill immune responses in Atlantic salmon following infection with *Neoparamoeba perurans*. *Frontiers in Immunology, 10*, 68.
Wang, Y., Wang, X., Lin, B., et al. (2019). Multi-omics approaches reveal molecular mechanisms of heat stress tolerance in tilapia. *Scientific Reports, 9*(1), Article 15781.
Wang, Y., Zheng, P., Peng, T., Yang, H., & Zou, J. (2020). Smart additive manufacturing: Current artificial intelligence-enabled methods and future perspectives. *Science China Technological Sciences, 63*(9), 1600–1611.
Wang, Z., & Srinivasan, R. S. (2017). A review of artificial intelligence based building energy use prediction: Contrasting the capabilities of single and ensemble prediction models. *Renewable and Sustainable Energy Reviews, 75*, 796–808.
Wishart, D. S. (2019). Metabolomics for investigating physiological and pathophysiological processes. *Physiological Reviews, 99*(4), 1819–1875.
Xu, C., Liu, D., Zhang, L., Xu, Z., He, W., Jiang, H., Zheng, M., & Qiao, N. (2021). AutoOmics: New multimodal approach for multi-omics research. *Artificial Intelligence in the Life Sciences, 1*, 100012.
Yáñez, J. M., Newman, S., & Houston, R. D. (2014). Genomics in aquaculture to better understand species biology and accelerate genetic progress. *Frontiers in Genetics, 5*, 302.
Yáñez, J. M., Newman, S., & Houston, R. D. (2015). Genomics in aquaculture to better understand species biology and accelerate genetic progress. *Frontiers in Genetics, 6*, 128.

Zhang, R., Shi, X., Liu, Z., Sun, J., Sun, T., & Lei, M. (2023). Histological, physiological and transcriptomic analysis reveal the acute alkalinity stress of the gill and hepatopancreas of *Litopenaeus vannamei*. *Marine Biotechnology, 25*(4), 588–602.

Zhang, S., Zheng, Y., Zhan, A., Dong, C., Zhao, J., & Yao, M. (2022). Environmental DNA captures native and non-native fish community variations across the lentic and lotic systems of a megacity. *Science Advances, 8*(6), eabk0097.

Zhang, X., et al. (2020). Proteomics advances in aquatic health: From stress response to biomarker discovery. *Aquaculture Reports, 18*, Article 100536.

Zhao, L., et al. (2021). Immune-related proteomic profiling in fish: Applications in vaccine development. *Fish & Shellfish Immunology, 113*, 10–22.

Zou, J., Huss, M., Abid, A., Mohammadi, P., Torkamani, A., & Telenti, A. (2019). A primer on deep learning in genomics. *Nature Genetics, 51*(1), 12–18.

11

Aquaculture drug design using AI and ML methods

Introduction

Aquaculture, the cultivation of aquatic animals and plants, including fish, crustaceans, mollusks, and algae, is the fastest-growing sector of the food industry worldwide. Aquaculture production has also had to contend with increasing disease infections that threaten productivity, animal health, and ecosystem balance (Martinez-Porchas & Martinez-Cordova, 2012). A series of pathogens, ranging from bacterial groups (e.g., *Aeromonas hydrophila*, Vibrio spp., *Edwardsiella tarda*) to viral pathogens (e.g., Infectious Salmon Anemia Virus, White Spot Syndrome Virus), and parasitic groups (e.g., *Ichthyophthirius multifiliis*, *Neoparamoeba perurans*), recurrently threaten farmed animals. These pathogens cause enormous financial losses each year, compromise supply chains, and frequently result in the extensive application of chemotherapeutics and antibiotics, with potential risks of drug resistance and environmental pollution (Okon et al., 2023). Traditional drug discovery for aquaculture is, by nature, slow, capital-consuming, and heavily based on trial and error. The drug development pipeline, from target identification to clinical proof, typically takes many years and is a costly endeavor involving significant investment in wet-lab experimentation, which may not be feasible for the rapidly changing disease dynamics in aquatic systems (Sun et al., 2022).

Additionally, aquatic organisms have different physiological and metabolic profiles, making most terrestrial animal medicines unsuitable. Lack of species-specific pharmacological data and the logistical challenges of drug delivery in aquatic environments also hinder standard approaches. There is thus a growing necessity to accelerate drug discovery pipelines specifically tailored to aquaculture's unique biological and environmental requirements (Hejna et al., 2022). The need for rapid, precise, and species-specific drug design has thus generated interest in the application of AI and ML technologies. These computational

Aquaculture with Artificial Intelligence. https://doi.org/10.1016/B978-0-443-33704-8.00014-8

methods hold the promise of transforming aquaculture pharmacology by facilitating predictive, high-throughput, and low-cost drug discovery and optimization. AI and ML algorithms can query intricate biological datasets, ranging from proteomics and genomics to pathogen-host interaction networks, and predict candidate drug targets and therapeutic activity with excellent accuracy. They enable virtual screening of compound libraries, de novo drug design, and in silico pharmacokinetics and toxicity profiling (Blanco-Gonzalez et al., 2023).

ML models, including deep neural networks, SVMs, and graph-based learning algorithms, were found to be very promising for the prediction of drug-pathogen interactions and optimization of compound structures toward binding specificity and affinity. In addition, generative models such as GANs and variational autoencoders (VAEs) are being increasingly utilized to design novel molecular structures with desirable pharmacological properties in a highly efficient manner, compared with the extensive chemical synthesis and biological testing required for traditional approaches. Through the incorporation of AI/ML methods into the drug design process, aquaculture scientists and pharmaceutical researchers can accelerate the identification of lead compounds, tailor treatments to various aquatic species, and respond quickly to newly emerging threats from diseases. Such an intelligent, data-driven approach to drug discovery is auspicious not only for enhancing the sustainability and resilience of aquaculture systems but also for mitigating the environmental threats associated with the untargeted application of drugs (Dara et al., 2022).

Key principles of drug design in aquatic species

The design of therapeutics for aquatic animals demands an entirely different approach than drug design for terrestrial animals, because the physiological, ecological, and environmental contexts in which these animals live are so different. Successful drug design in aquaculture should consider not only the pathogen–host interaction but also the pharmacokinetic (PK) and pharmacodynamic (PD) properties of drugs in aquatic animals, the complexities of drug delivery in waterborne systems, and the stringent regulatory systems for the approval and application of such therapeutics (Brodin et al., 2014). Pharmacokinetics (PK) refers to the processes of drug absorption, distribution, metabolism, and excretion (ADME), while pharmacodynamics (PD) relates to the drug's action, mechanism, and biological effects. PK/PD phenomena in aquatic animals depend on variables that differ significantly from those of terrestrial animals. For example, the poikilothermic (cold-blooded) physiology in fish and crustaceans results in temperature-dependent metabolic rates, which can impact drug absorption and clearance. Gill-based respiration and aquatic osmoregulation further impact how drugs are absorbed, particularly when administered through immersion or bath treatments.

Additionally, the presence of mucosal barriers and biofilms on skin and gill surfaces can alter the permeability and efficacy of drugs. These physiological differences necessitate species-specific modeling and careful dose adjustment to achieve therapeutic efficacy without causing toxicity or environmental contamination. Drug delivery and bioavailability are of utmost importance in aquaculture use. Oral and injectable routes of administration are most common in land animals, while aquatic animals typically require immersion, in-feed, or waterborne routes of administration. These pathways are susceptible to variation in drug absorption as a function of water temperature, pH, salinity, and the organism's stress response.

Additionally, the leaching of drugs into the surrounding water may lead to a loss of bioavailability, environmental contamination, and the development of microbial resistance. Targeted, efficient, and environmentally friendly delivery is arguably one of the most pressing concerns in the aquatic drug pipeline. State-of-the-art technologies, such as encapsulation, nanoparticle carriers, and computerized feed formulation, are being explored to maximize bioavailability and minimize waste (Okeke et al., 2022).

Production and use of aquaculture therapeutics are also tightly regulated. Regulatory bodies, such as the U.S. Food and Drug Administration (FDA) and the European Medicines Agency (EMA), regulate the approval process to ensure the safety of treated animals, consumers, and the environment. In the United States, the FDA's Center for Veterinary Medicine (CVM) reviews aquaculture medicines through the New Animal Drug Application (NADA) process, demanding comprehensive data on efficacy, target animal safety, human food safety, environmental effects, and manufacturing quality. The European regulatory framework requires similar criteria to be established through the Committee for Medicinal Products for Veterinary Use (CVMP). These regulatory phases, though critical, are both expensive and time-consuming, tending to discourage investment in medications for specialist or localized aquaculture species (Joppi et al., 2020). In this regard, AI and ML technologies offer solutions to major bottlenecks in aquatic drug development. Through the enablement of predictive PK/PD modeling, formulation optimization strategies, and evidence generation based on data to drive regulatory compliance, these technologies can reduce development pathways and costs while decreasing time to market (Malheiro et al., 2025).

Use of AI and ML in aquaculture drug discovery

The development of ML and AI is transforming the process of drug discovery in aquaculture into a data-driven process that is far superior to the conventional methods. These technologies enable rapid and precise target identification for treatment, intelligent screening and optimization of bioactive compounds, and realistic modeling of molecular interactions (Ferreira & Carneiro, 2025), all of which are critical in aquaculture, where species-specific information, environmental heterogeneity, and regulatory limitations pose unique challenges. This section addresses the multifaceted role of ML and AI in aquaculture pharmacology, from discovery and validation to target identification and ligand modeling (Fig. 11.1) (Ashraf Rather et al., 2024).

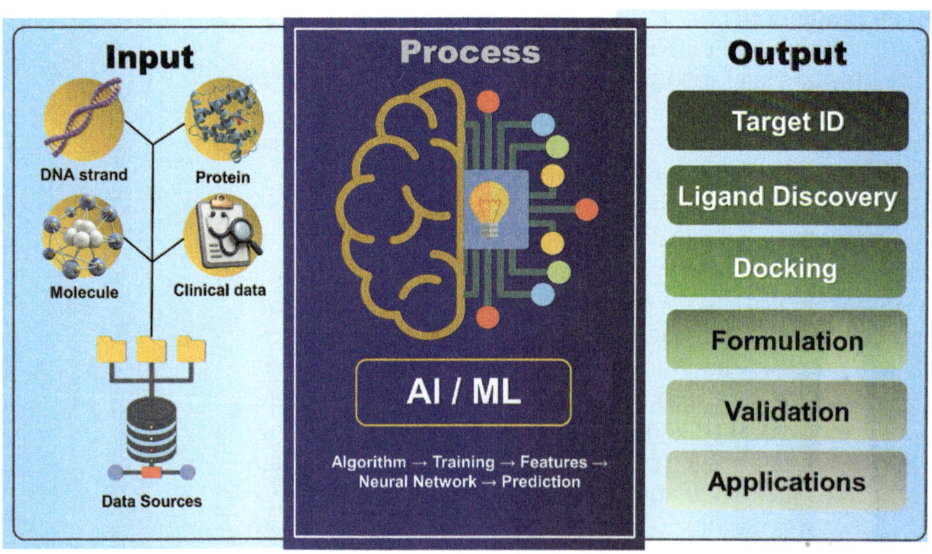

Figure 11.1 AI/ML input–process–output framework in aquaculture drug discovery, integrating multiomics, molecular data, and clinical records to generate predictive outputs.

Target identification and validation

The first and most fundamental step in drug discovery is the identification of molecular targets, such as receptors, enzymes, or genes that are essential to disease processes. AI and ML models, coupled with omics technologies (genomics, transcriptomics, proteomics, and metabolomics), provide an unprecedented understanding of pathogen-host interactions in aquatic animals. Omics data generated at high throughput are amenable to analysis by AI to identify differentially expressed proteins and genes in response to infection or stress. This information is graded into ML classifiers (Random Forest, SVMs) to rank putative druggable targets (Paananen & Fortino, 2020). Furthermore, AI-augmented pathway and network examination programs (graph-based algorithms, Bayesian networks, deep neural networks) may identify key nodes and interactions in critical biological networks responsible for immune response or pathogenicity. Genome-wide association studies (GWAS), increasingly applied to aquaculture species such as Atlantic salmon and tilapia, can also be assisted by ML models. These computer programs help identify associations between disease phenotypes and genetic variants in populations. AI-augmented GWAS tools not only enhance the precision of variant detection but also minimize false positives, making it easier to identify strong, high-confidence targets for therapeutic intervention (Caudai et al., 2021).

Ligand discovery and optimization

After the molecular targets are established as viable, the subsequent step is to discover and optimize ligand molecules that can bind to and modulate the target's activity. Conventional ligand discovery workflows often suffer from slow processing and are bottlenecked by the size of chemical libraries. Conversely, AI accelerates this with deep learning model-based virtual screening, such as DeepDock, AtomNet, and Graph Neural Networks (GraphNNs), which can screen millions of candidate molecules for binding affinity with high specificity. Quantitative structure–activity relationship (QSAR) modeling, which is one of the key ML approaches, is used to predict the biological activity of compounds from their chemical structure. Through the training of regression and classification models on current bioactivity databases, scientists can anticipate the behavior of a compound, facilitate informed decision-making and reducing the need for expensive wet-lab assays (Schenone et al., 2013).

Furthermore, de novo drug design through generative AI models, such as GANs and VAEs, is a revolutionary innovation in aquaculture drug discovery. These algorithms learn the distribution of chemical space and create completely novel molecules with optimized physicochemical and pharmacokinetic profiles. These models are especially valuable for drug design specific to aquatic species, where conventional compound libraries might be irrelevant or ineffective (Tang et al., 2024).

Molecular docking and dynamics simulations

A key component of drug efficacy assessment is a precise picture of how ligands bind to their targets at the molecular level. Molecular docking procedures supported by AI improve accuracy and processing speeds with intelligent scoring functions and conformational flexibility. Programs such as DeepDock, GNINA, and DeltaVina apply deep learning to predict binding pose and affinity with higher accuracy than classical algorithms. Historically, scoring functions have proven unsuitable for capturing the complexity of aquatic pathogen systems, particularly in the context of flexible proteins or metal ion coordination. Scoring functions based on ML, which are trained with extensive collections of protein–ligand complexes, can recognize nonlinear patterns of interaction and give more precise predictions of binding affinity, even in the presence of solvent effects or atypical active site topologies (Meng et al., 2011). Moreover, AI is also being used more extensively in molecular dynamics simulations, which are computationally demanding but necessary to learn the dynamic behavior of drug-target interactions over time. Force fields can be optimized using ML models by learning from quantum mechanical calculations so that more accurate energy calculations and conformational sampling can be done. AI is also employed in trajectory forecasting, simulation data anomaly detection, and simulation timelines acceleration with surrogate models (Zhang et al., 2022).

Data-driven approaches and resources

The foundation of AI- and ML-based drug discovery in aquaculture is the existence and effective utilization of high-quality, domain-specific data. Because aquatic species, pathogens, and environmental conditions are heterogeneous, data-driven approaches are needed to develop credible, accurate, and generalizable models. The success of such approaches, however, is irretrievably tied to the richness, quality, and contextual relevance of the data. Aquaculture pathogen databases are significant in providing curated biological and epidemiological data for a range of aquaculture diseases. The databases collect genomic data, virulence determinants, host-pathogen networks, and drug resistance in pathogens such as Vibrio spp., Aeromonas hydrophila, Streptococcus iniae, and other viral pathogens, including WSSV and ISAV. Their integration into AI workflows facilitates more precise target identification, comparative genomics, and prediction of aquaculture pathogen-specific resistance mechanisms. Integration of HTS data streamlines access to large-scale compound screening outputs, typically produced by biochemical or phenotypic assays. In silico models, utilizing HTS data as the training basis, can be employed to infer compound bioactivity, streamline hit-to-lead optimization, and minimize reliance on wet-lab validation on a large scale. ML models can identify concealed patterns in screening data and enhance candidate ranking for progression (Visan & Negut, 2024). Labeled datasets, in which compounds or targets are assigned biological results such as binding affinity, toxicity, or therapeutic potency, are necessary for supervised learning operations. These datasets form the basis for developing classification and regression models that can acquire significant structure–activity relationships and make precise predictions. In low-data settings, which are typical in aquaculture due to constraints on experimental resources, methods such as data augmentation (e.g., SMILES enumeration, structural perturbations) and transfer learning are invaluable. Transfer learning enables models that have been pretrained on large datasets (e.g., human or terrestrial animal drug data) to be fine-tuned on aquaculture-specific data, enhancing performance even when labeled examples are limited (Dou et al., 2023).

AI-enabled toxicity and resistance prediction

Ensuring the safety and sustainability of aquaculture therapeutics extends beyond efficacy; it must also encompass a comprehensive understanding of potential toxicity and environmental hazards. ML and AI models are now key components in forecasting adverse effects on both aquatic organisms and the environment, thereby facilitating risk assessment and aiding regulatory compliance. One of the primary challenges in drug design for aquaculture is predicting off-target effects and toxicity in fish and shellfish, which can vary significantly among species due to differences in metabolic pathways, gill permeability, and immune responses. Random forests, gradient boosting, and deep neural networks are some of the ML algorithms that can be trained on high-quality curated datasets of toxicity to predict acute and chronic toxicities. These models evaluate structural and physicochemical properties of compounds, including hydrophobicity, molecular weight, and electronic properties, to predict lethal concentration thresholds (e.g., LC50) and sublethal effects. The approach enables the early elimination of risky compounds, thereby reducing the need for costly in vivo testing (Zhou et al., 2023).

Along with target species safety, environmental impact, and degradation profiles are also predicted by ML models nowadays. By modeling physicochemical interactions with water, sediment, and microbes, AI can predict persistence, tendency toward bioaccumulation, and breakdown products of future drugs. This is essential to avoid long-term environmental disruption and to adhere to sustainable aquaculture practices, which is another urgent issue. Overuse and inadequate dosing in aquaculture contribute to the emergence of antibiotic resistance, which is further exacerbated by the nonscientific use of antibiotics (Evangelista et al., 2025). ML classifiers such as SVMs and ensemble methods are currently used on genetic and metagenomic sequences to detect resistance genes and predict trends in resistance under various treatment schemes. These technologies aid in maximizing antibiotic stewardship and creating alternative medicines, such as AI-designed antimicrobial peptides or phage therapies. In combination, these AI-based strategies for toxicity and resistance prediction enhance the safety and sustainability of aquaculture therapeutics, thereby contributing to the advancement of precision aquaculture and responsible drug development (Elalouf et al., 2025).

AI-driven formulation and delivery systems

Designing and delivering therapeutics in aquaculture presents special challenges, considering the aquatic environment, species physiology, and the need to avoid environmental contamination. AI is now revolutionizing the development of intelligent drug delivery systems with improved therapeutic effectiveness, accuracy, and environmental protection. One of the most promising applications is in intelligent feed formulation, which incorporates bioactive ingredients such as antibiotics, probiotics, immunostimulants, and functional peptides. AI models, specifically deep learning and optimization algorithms, can interpret complex nutritional and metabolic data to make predictions about the optimal feed composition for specific species, life stages, and health statuses. The models enable therapeutic agents to be incorporated into diets, ensuring consistent dosing, enhanced absorption, and minimal loss, thereby strengthening both animal health and environmental responsibility (Muteeb et al., 2023). AI is also transforming nanoparticle-based delivery system design for aquaculture. By learning from physicochemical data sets through ML, scientists can forecast the ideal size, surface charge, and material composition of nanoparticles to yield optimal drug encapsulation, stability, and release characteristics. AI-assisted nanocarriers deliver drugs to the site of action with increased bioavailability and a reduced dose requirement. Targeting can be achieved through pH-sensitive, enzyme-sensitive, or ligand-based systems that release the drug solely at infection sites or in target tissues, such as gastrointestinal tissue or gill epithelium (Chou et al., 2023).

Furthermore, AI-assisted modelling of release rates and penetration into bioparticles offers insight into what happens to formulations in the aqueous environment. ML codes are capable of modeling diffusion profiles and interactions with bio-barriers to predict sustained or controlled release profiles that match therapeutic windows. Furthermore, AI software packages are used to develop drugs that penetrate biofilms, which typically shield pathogens in aquaculture systems, thereby optimizing therapy for chronic conditions (Fig. 11.2) (Sarkar et al., 2023).

Case studies and applications

The incorporation of AI and ML in aquaculture drug discovery and health management has evolved from theoretical investigation into real-world application. Several case studies now demonstrate the practical benefits of AI-based tools in vaccine development, antibiotic optimization, enhanced disease surveillance, and precision dosing strategies for various aquaculture species. One of the most significant domains of application is AI-driven vaccine and antibiotic development against common pathogens, such as Aeromonas hydrophila, Vibrio spp., and Streptococcus iniae, which cause considerable mortality in finfish and shellfish. Scientists have utilized reverse vaccinology strategies driven by ML-based algorithms for the screening of

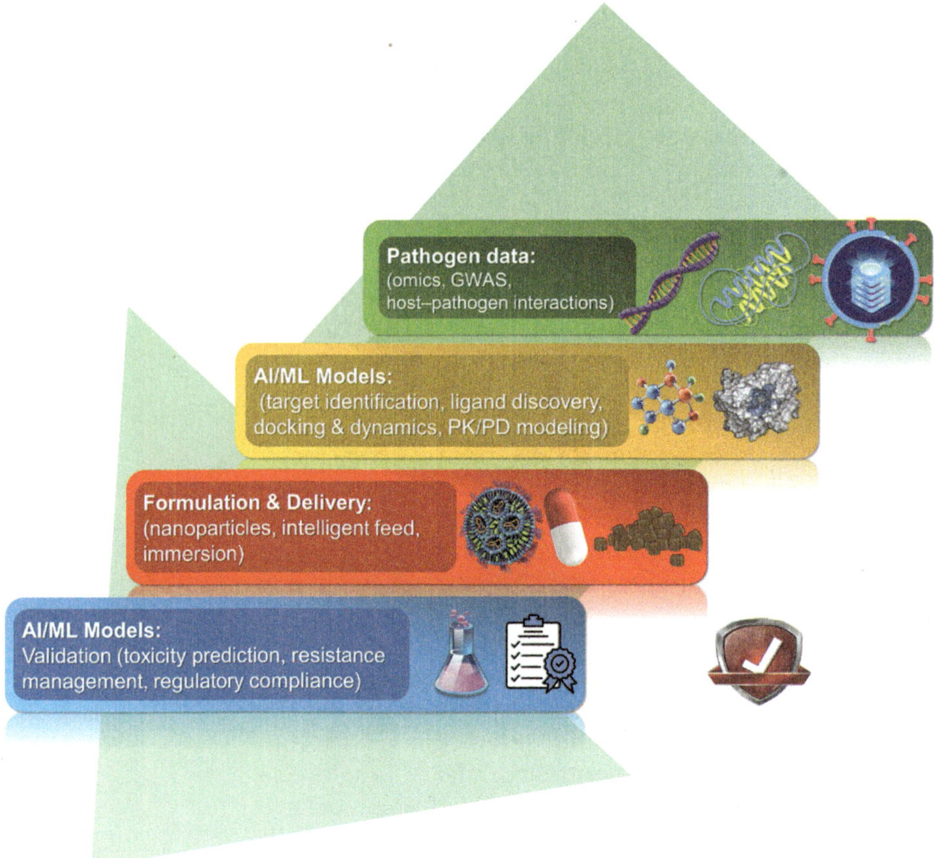

Figure 11.2 Workflow of AI/ML-driven aquaculture drug design, from pathogen data mining to formulation, delivery, and validation.

pathogen genomes to find conserved immunogenic epitopes that can be used as candidates for vaccine development. In silico models predict allergenicity, antigenicity, and stability at high speed, eliminating the need for wet-lab screening to filter out viable subunit vaccines. For instance, recent research has employed support vector machines and deep neural networks to select multiepitope peptide vaccines against Aeromonas and Vibrio, demonstrating encouraging immunogenicity in zebrafish and tilapia models. At the same time, ML models are also being developed in shrimp and salmon aquaculture to monitor health status and predict disease outbreaks (Serrano et al., 2024).

Deep learning and computer vision techniques, such as CNNs, are applied to underwater photography and video recordings for real-time behavioral monitoring, for example, swimming patterns or feeding behavior deviations, as preliminary indicators of infection or stress. In shrimp aquaculture, AI platforms combined with water quality sensors and microbiome information can predict the occurrence of White Spot Syndrome Virus outbreaks, facilitating timely intervention. In Atlantic salmon aquaculture, ML predictive tools have been utilized to accurately predict the occurrence of sea lice infestations and gill disease outbreaks, resulting in very low false positives, thereby enhancing biosecurity and reducing reliance on antibiotics (Voulodimos et al., 2018). One of the significant advancements in aquaculture therapeutics is the enhancement through precision dosing methods created through AI-based PK models. Current one-size-fits-all dosing protocols have the disadvantage of failing to account for interspecies variation, age of development, or environmental factors such as temperature and salinity, which can impact medication uptake and metabolism. PK models constructed using AI-aided historical treatment information, physiologic input, and external conditions can present individually tailored dosing regimens for optimizing therapy while preventing toxicity and resistance. As a specific example, AI-facilitated simulations have already been utilized in tilapia and catfish, and have computed the optimal oral antibiotic dosing interval to reduce medication intake by up to 30% without altering adequate plasma concentration levels (Muteeb et al., 2023). These practical applications illustrate the evolution of AI from a research curiosity to a foundation of intelligent aquaculture health management. Through the convergence of biological insight and computational accuracy, AI and ML tools are facilitating faster, more affordable, and species-specific interventions that enhance both productivity and sustainability. Additionally, the scalability and flexibility of such technologies position them well for use across a variety of aquaculture systems, ranging from land-based recirculating farms on an intensive scale to open-ocean cages and low-resource smallholder operations. With more data forthcoming and computational solutions improving, such case studies set the stage for a very encouraging future wherein aquaculture therapeutics facilitated through AI are capable of delivery to an unprecedented accuracy, speed, and sustainability basis (Fig. 11.3) (Bajwa et al., 2021).

Challenges and limitations

Despite the encouraging progress in AI and ML for aquaculture drug discovery, several challenges and limitations prevent

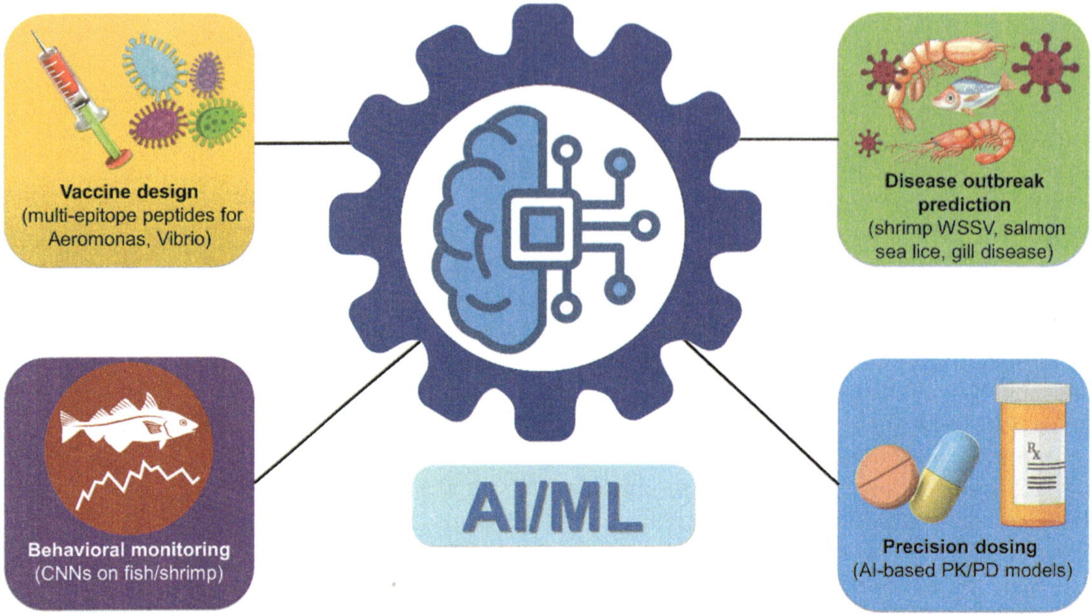

Figure 11.3 Practical application of AI and ML in aquaculture therapeutics and health management.

their general application and translational success. One key hindrance is data paucity, especially for nonmodel aquatic species and region-specific pathogens. Accurate, generalizable AI models require high-quality, labeled datasets; however, the aquaculture industry often lacks thorough genomic, phenotypic, and pharmacological datasets for most commercially important species. Additionally, species-to-species variability in physiology, immunity, and metabolism makes model development particularly challenging. Models trained on one species or condition can yield poor results when transferred to another species or condition, underscoring the need for species-specific, context-aware models and transfer learning approaches (Blanco-Gonzalez et al., 2023). A second significant limitation is model interpretability and real-world validation. Most high-performing AI models, particularly deep neural networks, are "black boxes" with limited transparency in their rationale for predictions. In a domain such as aquaculture, where therapeutic decisions directly impact animal health, food safety, and ecosystem integrity, model transparency is crucial. Moreover, many AI models are trained and tested in controlled in silico or laboratory environments, and their performance tends to deteriorate when applied in actual aquaculture environments with dynamic biological, ecological, and operational factors. Field-based validation and the inclusion of uncertainty quantification in model predictions are still urgently needed (Cembrowska-Lech et al., 2023). Finally, ethical, ecological, and regulatory issues need to be addressed. The use of AI-suggested therapeutics without a thorough safety assessment could introduce ecological hazards, including bioaccumulation, species interactions, or antimicrobial resistance. Regulatory bodies must adapt to accommodate the incorporation of AI-based drug development pipelines, striking a balance between innovation and stringent regulation. Additionally, ethical issues related to data ownership, algorithm bias, and socioeconomic implications for smallholder farmers must be addressed in an appropriate manner (Mennella et al., 2024).

References

Ashraf Rather, M., Ahmad, I., Shah, A., Ahmad Hajam, Y., Amin, A., Khursheed, S., … Rasool, S. (2024). Exploring opportunities of artificial intelligence in aquaculture to meet increasing food demand. *Food Chemistry X, 22*, Article 101309. https://doi.org/10.1016/j.fochx.2024.101309

Bajwa, J., Munir, U., Nori, A., & Williams, B. (2021). Artificial intelligence in healthcare: transforming the practice of medicine. *Future Healthcare Journal, 8*(2), e188–e194. https://doi.org/10.7861/fhj.2021-0095

Blanco-Gonzalez, A., Cabezon, A., Seco-Gonzalez, A., Conde-Torres, D., Antelo-Riveiro, P., Pineiro, A., & Garcia-Fandino, R. (2023). The role of AI in drug discovery: challenges, opportunities, and strategies. *Pharmaceuticals, 16*(6). https://doi.org/10.3390/ph16060891

Brodin, T., Piovano, S., Fick, J., Klaminder, J., Heynen, M., & Jonsson, M. (2014). Ecological effects of pharmaceuticals in aquatic systems—impacts through behavioural alterations. *Philosophical Transactions of the Royal Society of London B Biological Sciences, 369*(1656). https://doi.org/10.1098/rstb.2013.0580

Caudai, C., Galizia, A., Geraci, F., Le Pera, L., Morea, V., Salerno, E., … Colombo, T. (2021). AI applications in functional genomics. *Computational and Structural Biotechnology Journal, 19*, 5762–5790. https://doi.org/10.1016/j.csbj.2021.10.009

Cembrowska-Lech, D., Krzeminska, A., Miller, T., Nowakowska, A., Adamski, C., Radaczynska, M., … Mikiciuk, M. (2023). An integrated multi-omics and artificial intelligence framework for advance plant phenotyping in horticulture. *Biology, 12*(10). https://doi.org/10.3390/biology12101298

Chou, W. C., Chen, Q., Yuan, L., Cheng, Y. H., He, C., Monteiro-Riviere, N. A., … Lin, Z. (2023). An artificial intelligence-assisted physiologically-based pharmacokinetic model to predict nanoparticle delivery to tumors in mice. *Journal of Controlled Release, 361*, 53–63. https://doi.org/10.1016/j.jconrel.2023.07.040

Dara, S., Dhamercherla, S., Jadav, S. S., Babu, C. M., & Ahsan, M. J. (2022). Machine learning in drug discovery: a review. *Artificial Intelligence Review, 55*(3), 1947–1999. https://doi.org/10.1007/s10462-021-10058-4

Dou, B., Zhu, Z., Merkurjev, E., Ke, L., Chen, L., Jiang, J., … Wei, G. W. (2023). Machine learning methods for small data challenges in molecular science.

Chemical Reviews, 123(13), 8736–8780. https://doi.org/10.1021/acs.chemrev.3c00189

Elalouf, A., Elalouf, H., Rosenfeld, A., & Maoz, H. (2025). Artificial intelligence in drug resistance management. *3 Biotech, 15*(5), 126. https://doi.org/10.1007/s13205-025-04282-w

Evangelista, D., Nelson, E., Skyner, R., Tehan, B., Bernetti, M., Roberti, M., ... Bottegoni, G. (2025). Application of deep learning to predict the persistence, bioaccumulation, and toxicity of pharmaceuticals. *Journal of Chemical Information and Modeling, 65*(7), 3248–3261. https://doi.org/10.1021/acs.jcim.4c02293

Ferreira, F. J. N., & Carneiro, A. S. (2025). AI-Driven drug discovery: a comprehensive review. *ACS Omega, 10*(23), 23889–23903. https://doi.org/10.1021/acsomega.5c00549

Hejna, M., Kapuscinska, D., & Aksmann, A. (2022). Pharmaceuticals in the aquatic environment: a review on eco-toxicology and the remediation potential of algae. *International Journal of Environmental Research and Public Health, 19*(13). https://doi.org/10.3390/ijerph19137717

Joppi, R., Bertele, V., Vannini, T., Garattini, S., & Banzi, R. (2020). Food and drug administration vs European medicines agency: review times and clinical evidence on novel drugs at the time of approval. *British Journal of Clinical Pharmacology, 86*(1), 170–174. https://doi.org/10.1111/bcp.14130

Malheiro, V., Santos, B., Figueiras, A., & Mascarenhas-Melo, F. (2025). The potential of artificial intelligence in pharmaceutical innovation: from drug discovery to clinical trials. *Pharmaceuticals, 18*(6). https://doi.org/10.3390/ph18060788

Martinez-Porchas, M., & Martinez-Cordova, L. R. (2012). World aquaculture: environmental impacts and troubleshooting alternatives. *The Scientific World Journal, 2012*, Article 389623. https://doi.org/10.1100/2012/389623

Meng, X. Y., Zhang, H. X., Mezei, M., & Cui, M. (2011). Molecular docking: a powerful approach for structure-based drug discovery. *Current Computer-Aided Drug Design, 7*(2), 146–157. https://doi.org/10.2174/157340911795677602

Mennella, C., Maniscalco, U., De Pietro, G., & Esposito, M. (2024). Ethical and regulatory challenges of AI technologies in healthcare: a narrative review. *Heliyon, 10*(4), Article e26297. https://doi.org/10.1016/j.heliyon.2024.e26297

Muteeb, G., Rehman, M. T., Shahwan, M., & Aatif, M. (2023). Origin of antibiotics and antibiotic resistance, and their impacts on drug development: a narrative review. *Pharmaceuticals, 16*(11). https://doi.org/10.3390/ph16111615

Okeke, E. S., Chukwudozie, K. I., Nyaruaba, R., Ita, R. E., Oladipo, A., Ejeromedoghene, O., ... Okoye, C. O. (2022). Antibiotic resistance in aquaculture and aquatic organisms: a review of current nanotechnology applications for sustainable management. *Environmental Science and Pollution Research International, 29*(46), 69241–69274. https://doi.org/10.1007/s11356-022-22319-y

Okon, E. M., Okocha, R. C., Taiwo, A. B., Michael, F. B., & Bolanle, A. M. (2023). Dynamics of co-infection in fish: a review of pathogen-host interaction and clinical outcome. *Fish and Shellfish Immunology Reports, 4*, Article 100096. https://doi.org/10.1016/j.fsirep.2023.100096

Paananen, J., & Fortino, V. (2020). An omics perspective on drug target discovery platforms. *Briefings in Bioinformatics, 21*(6), 1937–1953. https://doi.org/10.1093/bib/bbz122

Sarkar, C., Das, B., Rawat, V. S., Wahlang, J. B., Nongpiur, A., Tiewsoh, I., … Sony, H. T. (2023). Artificial intelligence and machine learning technology driven modern drug discovery and development. *International Journal of Molecular Sciences, 24*(3). https://doi.org/10.3390/ijms24032026

Schenone, M., Dancik, V., Wagner, B. K., & Clemons, P. A. (2013). Target identification and mechanism of action in chemical biology and drug discovery. *Nature Chemical Biology, 9*(4), 232–240. https://doi.org/10.1038/nchembio.1199

Serrano, D. R., Luciano, F. C., Anaya, B. J., Ongoren, B., Kara, A., Molina, G., … Lalatsa, A. (2024). Artificial intelligence (AI) applications in drug discovery and drug delivery: revolutionizing personalized medicine. *Pharmaceutics, 16*(10). https://doi.org/10.3390/pharmaceutics16101328

Sun, D., Gao, W., Hu, H., & Zhou, S. (2022). Why 90% of clinical drug development fails and how to improve it? *Acta Pharmaceutica Sinica B, 12*(7), 3049–3062. https://doi.org/10.1016/j.apsb.2022.02.002

Tang, X., Dai, H., Knight, E., Wu, F., Li, Y., Li, T., & Gerstein, M. (2024). A survey of generative AI for de novo drug design: new frontiers in molecule and protein generation. *Briefings in Bioinformatics, 25*(4). https://doi.org/10.1093/bib/bbae338

Visan, A. I., & Negut, I. (2024). Integrating artificial intelligence for drug discovery in the context of revolutionizing drug delivery. *Life, 14*(2). https://doi.org/10.3390/life14020233

Voulodimos, A., Doulamis, N., Doulamis, A., & Protopapadakis, E. (2018). Deep learning for computer vision: a brief review. *Computational Intelligence and Neuroscience, 2018*, Article 7068349. https://doi.org/10.1155/2018/7068349

Zhang, Y., Luo, M., Wu, P., Wu, S., Lee, T. Y., & Bai, C. (2022). Application of computational biology and artificial intelligence in drug design. *International Journal of Molecular Sciences, 23*(21). https://doi.org/10.3390/ijms232113568

Zhou, Y., Wang, Y., Peijnenburg, W., Vijver, M. G., Balraadjsing, S., & Fan, W. (2023). Using machine learning to predict adverse effects of metallic nanomaterials to various aquatic organisms. *Environmental Science & Technology, 57*(46), 17786–17795. https://doi.org/10.1021/acs.est.2c07039

12

Blockchain, cloud computing, and big data for aquaculture

Aspirants aquaculture

The incorporation of AI and ML in aquaculture is still in its early stages; yet, the pace of technology development holds promise that more innovative, self-contained, and highly responsive agricultural systems will become the norm. With the expansion of the sector toward more efficient and sustainable production, some emerging patterns have the potential to reinvent the construction of AI and ML models and how they are deployed, adopted, and understood within aquaculture systems. One of the most significant breakthroughs is the emergence of Explainable AI (XAI), which addresses the "black box" approach of many sophisticated models. For aquaculture, where decisions directly impact animal welfare, productivity, and sustainability, transparency in decision-making is paramount. XAI methods aim to enhance the interpretation of AI systems by providing clear, transparent explanations of model predictions, thereby fostering trust among farmers, policymakers, and other stakeholders. This openness is necessary to promote the use of AI-based solutions in traditionally conservative aquaculture farms (Ashraf Rather et al., 2024). Federated Learning is another trend that revolutionizes, a technique that allows AI models to be trained across multiple decentralized farms without explicit data sharing. With the sensitivity and proprietary nature of aquaculture data, ranging from breeding records to farm-specific environmental parameters, federated learning offers a way to collaboratively develop models without sacrificing data privacy. This strategy can enable the creation of more general and robust models by leveraging different sets of data across various geographical and environmental settings (Zalik & Zalik, 2023). IoT and edge computing technologies are also being increasingly utilized in conjunction with each other. By placing smart sensors and computational power within the location of data collection itself, edge computing mitigates latency, supports real-time

Aquaculture with Artificial Intelligence. https://doi.org/10.1016/B978-0-443-33704-8.00012-4

analysis, and reduces the need for centralized cloud-based infrastructures. In aquaculture, where a quick response to environmental shifts or disease is crucial, such facilities offer tremendous operational advantages. IoT-based systems can continuously monitor water quality, feeding behavior, and fish health and input the data directly into AI models for instant action (Nguyen et al., 2024). Finally, the use of Generative AI promises to simulate intricate aquaculture situations. Synthetic data for infrequent events (e.g., illnesses) can be generated by generative models, such as generative adversarial networks (GANs) or diffusion models, to mimic environmental disturbances and thereby enable the training and verification of AI systems under various scenarios. Such capability not only surmounts the lack of authentic data but also allows the design of responsive aquaculture systems that can manage future climatic and ecological uncertainties (Arora & Arora, 2022).

Blockchain

The global aquaculture industry, the mainstay of economic and food security development, is being transformed with digital innovation. With the escalation of the size and complexity of aquaculture business operations, traditional practices that rely on manual monitoring and reactive management do not sufficiently meet the challenges of achieving higher productivity, environmental sustainability, and food safety. Thus, the digitalization of aquaculture has become a crucial agenda item to maximize efficiencies, reduce ecological footprints, and promote transparency throughout the value chain. Frontier technologies, such as blockchain, cloud computing, and big data analytics, are increasingly recognized as key enablers of this transformation. These technologies not only enhance the production efficiency of aquaculture farms but also improve traceability, accountability, and sustainability, which are crucial parameters at a time of increased consumer anxiety, stringent regulatory demands, and climatic pressures. Digital technologies enable farmers to transition from experience-based management to evidence-based, forward-looking models of decision-making, thereby shaping the future of aquaculture (Pradeepkiran, 2019). Blockchain technology offers a secure distributed ledger system to demonstrate data integrity and traceability throughout the aquaculture supply chain. From hatchery to harvest and retail distribution, blockchain can provide transparent records of origin, certification of health, feeding regimes, and environmental conditions, thereby enhancing compliance and consumer confidence.

Furthermore, computer programs on the blockchain, known as smart contracts, can be used to automate central transactions, such as stock damage claims or paying suppliers, thereby reducing the costs of back-office operations and enhancing operational flexibility (Patel et al., 2023). Cloud computing plays a central role in facilitating the real-time aggregation, storage, and computation of vast volumes of data. As more and more IoT sensors measure water quality parameters, biomass estimates, feeding rates, and behavior, cloud platforms provide the computational power and scalability to process such vast volumes of data. Cloud solutions provide remote access, facilitate the straightforward integration of disparate data sources, and offer robust disaster recovery, enabling more resilient and collaborative aquaculture systems (Thilakarathne et al., 2025). Big data analytics also uncovers the potential of this digital space by converting raw operational and sensor data into actionable intelligence. Through predictive modeling, anomaly detection, and trend analysis, big data software empowers farm managers to make informed decisions on feeding habits, disease management, resource allocation, and environmental preservation. Advanced analytical techniques such as ML techniques and real-time visualization dashboards enable precision aquaculture that optimizes production while minimizing footprint. The triad of blockchain, cloud computing, and big data has the potential to disrupt the future of aquaculture. The three technologies combined can create a highly connected, transparent, and intelligent ecosystem where data flows securely and seamlessly among farm operations, supply chain participants, and regulatory bodies. For example, sensor data gathered in real-time on the farm can be safely stored in the cloud, processed using advanced data methods for predictive analysis, and verified with blockchain to ensure data integrity and traceability. Such a combined digital infrastructure not only enhances operational efficiency but also enhances the credibility of aquaculture products in international markets, facilitates certification efforts, and promotes sustainable practices (Hussein et al., 2025).

Big data in aquaculture

Big data technology is being widely adopted in aquaculture to transform the industry into a more informed, efficient, and sustainable enterprise. As aquaculture production spreads across geographies to meet the world's seafood needs, data at various production levels have multiplied in size and complexity manifold. Utilizing this knowledge in conjunction with advanced analytics has unrivaled potential to maximize productivity, enhance animal well-being and health, promote environmental sustainability, and optimize profitability throughout the entire aquaculture value chain. Identifying the sources, types, and applications used for big data in aquaculture is essential to maximize potential in this digital age (Bunting et al., 2023).

Sources of big data

Aquaculture big data originates from a variety of typically interconnected sources that capture the biological, environmental, operational, and economic aspects of farm systems. Ecological data are crucial for aquaculture management. Physical parameters, such as water temperature, salinity, dissolved oxygen levels, pH, turbidity, and nutrient concentrations, are continuously monitored through IoT-enabled sensors, autonomous underwater vehicles, and satellite remote sensing (Ahmed et al., 2024). These environmental factors have a direct impact on fish health, growth rates, and disease susceptibility; therefore, their real-time monitoring is crucial for adaptive farm management. Biological data is another key source. It encompasses data on fish growth rates, survival rates, feeding, reproductive performance, health status, and disease incidence. Wearable sensors, vision systems, and acoustics technologies are being used increasingly to monitor and record these biological parameters at high resolutions for early disease detection, behavioral insight, and individualized stock management (Hegde et al., 2023). Operations data measure the management routines adopted at aquaculture farms. Computerized farm management systems provide data about feeding schedules, feed conversion ratio, harvest dates, stock movement, equipment operation, and maintenance schedules on a day-to-day basis. Information of this type supports day-to-day optimization and long-term planning. Supply chain and economic data also come to the forefront. It consists of data related to production costs, sales data, market price data, transportation schedules, product traceability, and compliance document data. Combining production data with economic data enables the development of end-to-end management frameworks that balance fiscal outcomes and biological productivity (Thilakarathne et al., 2025).

Characteristics of big data in aquaculture

Big data in aquaculture has the typical features:

1. Volume: Modern aquaculture farms generate terabytes of data daily from thousands of sensors, video streams, and operational logs (Wang et al., 2021).
2. Diversity: The data varies in nature, ranging from quantitative sensor measurements and images to unstructured text reports and documents (Sedlakova et al., 2023).
3. Velocity: Data is generated at high velocities, necessitating real-time processing for actionable insights, particularly in critical environmental monitoring applications (Ramadan et al., 2024).
4. Truthfulness: Maintaining accurate and intact data is challenging, especially with sensor drift, missing data, and aquatic source noise (Mohd Jais et al., 2024).
5. Value: The Indispensable value of this data lies in its utility for facilitating farm operations, reducing risk, and enhancing profitability (Wangen et al., 2021).

Secondly, aquaculture big data encompasses both real-time data and historical data. Real-time data enables real-time decision-making, such as adjusting the feeding rate or responding to a lack of oxygen. In contrast, historical data serves as a basis for trend analysis, model training, and long-term planning. Organizationally, aquaculture data can be:

6. Structured data, e.g., sensor data and monetary data in relational databases.
7. Partially structured data, including JSON logs from IoT devices.
8. Unstructured data, such as underwater video data, can be applied for behavior observation or biomass measurement.

Big Data Analytics requires advanced data storage, efficient data preprocessing, and flexible analytical models for handling and processing these varied kinds of data (Sivarajah et al., 2017).

Applications of big data analytics

Strategic use of big data analytics in aquaculture is driving substantive innovation in multiple areas of operations. Preventive predictive disease outbreak is perhaps one of the most significant uses. Predictive models can forecast disease outbreaks before clinical presentation by utilizing historical trends and real-time biological and environmental data. Predictive warnings, coupled with early interventions, will prevent mass death and minimize the consumption of antibiotics, while enhancing animal welfare and the environment (Corsi et al., 2021). Resource optimization and feed management are other significant applications. Feed costs account for a substantial portion of operating expenses in aquaculture. Through real-time monitoring of feeding behaviors, climatic conditions, and fish biomass, big data analytics can control feeding times and volumes to maximize FCR with minimal waste. Similarly, data-driven optimizations of aeration, filtration, and energy consumption contribute to achieving resource-friendly farming (Ashraf Rather et al., 2024). Precision farming, driven by actionable information, is rapidly becoming the standard. Advanced analytics enable the customized management of various cages, ponds, or tanks within a single facility. By identifying microvariations in climatic factors and fish health, farmers can implement targeted interventions that enhance overall farm productivity and sustainability. Financial preparation and risk evaluation are also greatly facilitated through big data analytics. By linking economic, biological, and operational metrics, farmers and investors can better position themselves to evaluate risks associated with disease outbreaks, market fluctuations, and climatic variations. Predictive budget models enable more accurate budgeting, investment planning, and insurance premium estimation, thereby strengthening economic resilience across the aquaculture industry (Fig. 12.1) (Getahun et al., 2024).

Cloud computing in aquaculture

Cloud infrastructure for data management

The emergence of digital technologies in aquaculture has generated a substantial amount of data from various sources, including IoT sensors, feeders, cameras, and biological monitoring devices. All this information needs to be processed by a high-capacity, expandable, and secure infrastructure, and cloud computing has been the cornerstone of modern aquaculture data management (Rejeb et al., 2025). Remote data storage and backup are among the most fundamental advantages offered by cloud computing. Instead of relying on localized servers that are vulnerable to physical damage, data loss, or theft, aquaculture farms can utilize distributed cloud servers to store critical information such as historical water quality records, stock performance data, feed optimization reports, and environmental logs. This ensures data persistence, supports speedy disaster recovery in the event of system failure, and enables safe backup

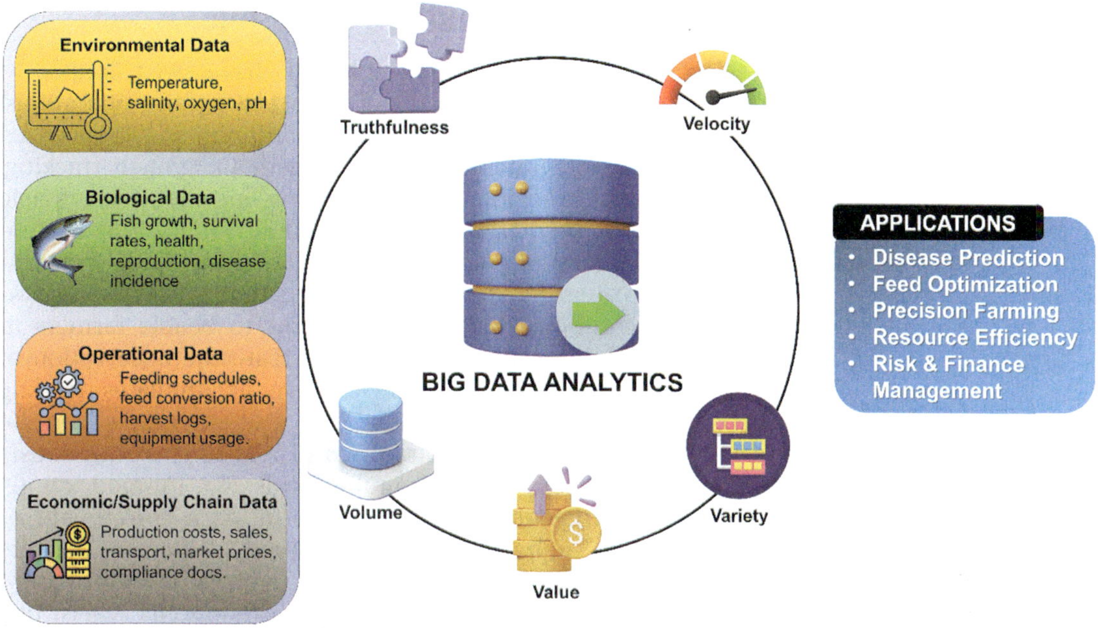

Figure 12.1 Big data in aquaculture: Sources, characteristics & applications.

procedures, which are a necessity when adhering to regulatory data retention requirements (Neethirajan, 2025). Scalability in the cloud presents a significant boon for aquaculture ventures that utilize large sensor arrays in scale- and resolution-based high-definition video surveillance systems. While sensor networks grow in number and image sensors consume terabytes of environmental or behavioral video, cloud infrastructure can dynamically allocate storage and compute capacity to handle such surges without needing to install on-premises, costly infrastructure. Such elasticity enables farms to scale efficiently, providing the data collection and analytics capacity required to meet the needs of both smallholder farms and large, multisite operations (Casino et al., 2025). Access from various locations of a farm also enhances operating flexibility. Cloud technology enables farm managers, aquaculture specialists, and veterinary consultants to analyze real-time and historical data from geographically dispersed locations using safe, web-enabled devices. This supports centralized decision-making, facilitates the remote management of offshore cages and terrestrial ponds, and promotes collaboration among multistakeholder teams geographically dispersed across various areas (Rowan et al., 2022).

Cloud-based aquaculture platforms

Several cloud-based expert solutions have been created to meet the specific requirements of aquaculture farm operations. Aqua Manager, for example, offers a combined cloud-based solution that integrates feeding management, growth tracking, environmental tracking, and harvesting logistics in a single dashboard. Its predictive analytics modules utilize past data to optimize production planning, reduce FCR, and lower environmental footprints. Similarly, fishery offers innovative feeding systems that combine cloud-based services to enable farmers to automate feeding schedules based on real-time observation of fish populations' behavior. The system accumulates feeding behavior information from IoT sensors and aggregates it in the cloud to optimize feeding strategies in real-time, continuously enhancing efficiency with minimal wastage (Ahmed et al., 2024). IoT connection to cloud platforms is a component of contemporary aquaculture management systems. Sensors installed in ponds, tanks, and offshore farms gather information regarding temperature, dissolved oxygen, ammonia levels, pH, and fish activity. Real-time data is sent to cloud servers, which are aggregated, processed, and analyzed through advanced ML techniques. IoT integration enables proactive responses to environmental changes, early detection of disease outbreaks, and optimized resource utilization based on predictive patterns. Cloud platforms also include real-time dashboards that provide intuitive graphical representations of key performance indicators (KPIs), such as growth rates, mortality events, water quality over time, and operational alerts. These dashboards enable farm managers to make instant decisions, adding to the timeliness and reducing the opportunity for rapid loss due to climatic or biological disturbances (Fig. 12.2) (Chen et al., 2022).

Blockchain for aquaculture

The aquaculture sector, increasingly beleaguered by issues related to sustainability, transparency, and food safety, is starting to adopt blockchain technology as a game-changing digital solution. Blockchain utilizes a decentralized, tamper-evident system of ledgers that can enhance traceability, facilitate food certification processes, safeguard supply chains, and automate transactions. Its use in aquaculture operations has the potential to revolutionize how data is transmitted, validated, and used in the farm-to-fork supply chain (Pakseresht et al., 2024).

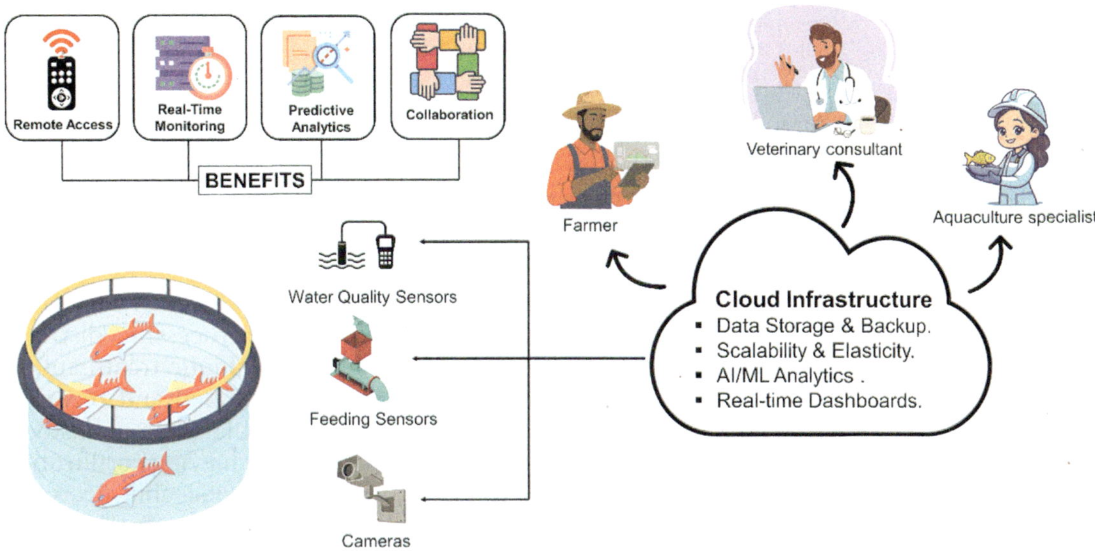

Figure 12.2 Cloud & IoT integration for smart aquaculture management.

Fundamentals of blockchain technology

In essence, blockchain is a type of distributed ledger technology in which transaction records are stored by a cluster of computers, or nodes, rather than a single central server. A list of transactions, dated and time-stamped, is contained in each block and is linked to the previous block cryptographically, ensuring that data cannot be manipulated retroactively once stored without network consensus. This immutable and open system confers immense advantages on aquaculture, where auditable records are essential for regulatory compliance, consumer trust, and operational integrity (Musa et al., 2023). One of the breakthroughs in blockchain systems is the use of smart contracts, computer programs that automatically execute the terms of a contract when certain conditions are met. In aquaculture, smart contracts can facilitate several automated transactions, such as releasing insurance payouts when environmental sensors indicate disastrous degradation of water quality or paying suppliers on verified delivery of feedstock or equipment. Blockchains are categorized into two types: permissionless and permissioned systems. Permissionless blockchains (e.g., Bitcoin, Ethereum) are open networks in which anyone can join without permission. Permissioned blockchains, in contrast, limit access to prevetted participants and are better suited to industries such as aquaculture, where regulations, privacy considerations, and governance need mandate-controlled access. Hyperledger Fabric and Quorum are permissioned blockchain platforms widely used in industrial environments (Khan et al., 2021).

Applications of blockchain in aquaculture

Traceability: From hatchery to plate with fish

Blockchain can create an indelible digital record of a fish's life cycle, from hatchery to farming, harvesting, processing, distribution, and retail. Any event, vaccination, antibiotic treatment, or environmental monitoring can be time-stamped and recorded on the blockchain, forming a secure, verifiable chain of custody. Such traceability not only assists farms in achieving certification but also enables consumers to make knowledgeable decisions about the source and sustainability of their seafood (Ismail et al., 2023).

Food safety: Verifying disease-free, sustainably produced seafood

Blockchain certification systems offer a robust platform for enhancing food safety in aquaculture. Pathogen testing results or water quality measurements that are critical control points can be documented immutably. In the event of a food safety crisis, blockchain facilitates quick and accurate trace-back to the origin, keeping recalls small and maintaining brand reputation. Additionally, sustainability certifications (such as organic standards and eco-labels) can be openly governed on blockchain networks, thereby improving confidence in product claims (Patel et al., 2023).

Supply chain transparency: Ensuring fair trade and authenticity

Global seafood supply chains are notoriously complex and susceptible to fraud, including mislabeling and labor abuses. Blockchain can increase supply chain transparency by recording every transaction and movement of aquaculture products. Stakeholders, farmers, processors, distributors, retailers, and consumers can verify each step in real time. This openness facilitates ethical sourcing, promotes fair trade, and enhances the fight against illegal, unreported, and unregulated (IUU) fishing practices, which often manipulate global seafood markets (Luna et al., 2024).

Data integrity: Avoiding data tampering and assuring auditability

In aquaculture, water quality data, feed input, growth rates, and harvest records are essential for management and external audits. By recording this data on a blockchain, farms can prevent tampering, human error, and post-hoc data manipulation. Immutable records provide a trustworthy audit trail that can be presented to regulators, investors, and certification bodies, improving operational credibility and access to premium markets (Ullah et al., 2024).

Smart contracts: Transacting automatically

Smart contracts provide a further level of automation and efficiency to aquaculture operations. Some examples include parametric insurance smart contracts that automatically make payments based on blockchain-verifiable sensor data reports of catastrophic events, such as toxic algal blooms or mass mortality events. Similarly, procurement systems may be automated: once feed stocks drop below a certain blockchain-confirmed level, a smart contract might automatically order from a preapproved supplier, saving administrative time and maintaining business Continuity (Fig. 12.3) (Pranto et al., 2021).

Case studies and pilot projects

Several pioneering projects have already shown the promise of blockchain for aquaculture and seafood supply chains. The IBM Food Trust platform is one such initiative that has successfully integrated blockchain technology to enhance traceability in the food industry, including the seafood sector. By allowing producers, processors, and retailers to upload critical information at every step of the supply chain, IBM Food Trust provides a transparent and secure flow of information, enabling rapid recalls, verifying product authenticity, and increasing consumer confidence (Pakseresht et al., 2024). Another is Provenance.org, a technology platform that allows companies to establish trust in their supply chains through blockchain. In seafood, Provenance.org

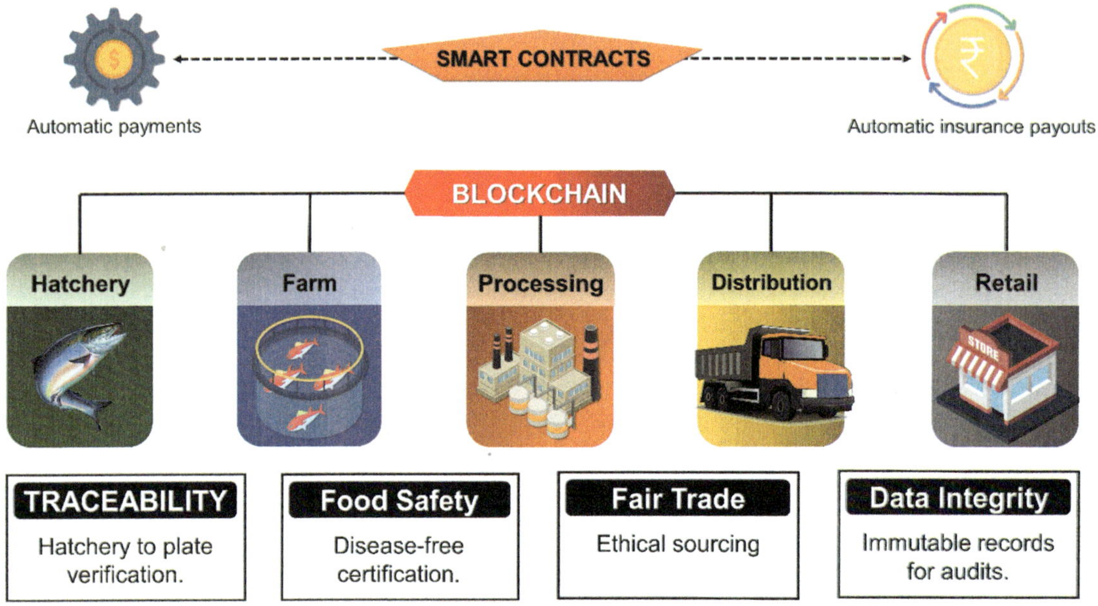

Figure 12.3 Blockchain for aquaculture supply chain transparency and automation.

has developed programs that enable consumers to scan QR codes on fish packaging and immediately receive verified information about the fish's journey, from catch to counter, including sustainability certifications and the labor practices behind its production. Furthermore, startup ecosystems and NGOs are collaborating on blockchain pilots for fisheries and aquaculture in Southeast Asia, where issues such as overfishing, mislabeling, and human rights abuses are particularly acute. These pilot projects demonstrate blockchain's capacity not only to protect ecological resources but also to improve livelihoods and governance in vulnerable coastal communities (Cromwell et al., 2025).

Integration of blockchain, cloud, and big data

The convergence of blockchain, cloud computing, and big data analytics represents a paradigm shift toward digitalization in aquaculture, enabling the creation of innovative, open, and highly reliable farming ecosystems. By converging these technologies, aquaculture businesses can achieve end-to-end visibility, data-driven decision-making, and improved sustainability, thereby solving many of the traditional problems that have historically hindered productivity and traceability in the sector. Underlying such integration is the vision to develop an intelligent and clear aquaculture system. Blockchain technology ensures that all data points, whether related to water quality characteristics, food consumption, health status, or harvesting operations, are stored safely and in an immutable manner. Such an immutable ledger enables traceability throughout the aquaculture supply chain, thereby reinforcing consumer confidence, ensuring regulatory compliance, and promoting operational accountability. Simultaneously, big data analytics platforms compile and analyze large amounts of sensor and operational data, deriving actionable insights to optimize farming operations and predict issues such as disease outbreaks or machinery breakdowns. Interoperability between blockchain networks and big data analytics platforms is needed to achieve the full potential of this combination. Analytics platforms translate and process the sophisticated data sets created at the farm level, and blockchain offers a way to validate and certify such insights (Rowan et al., 2022). For example, an early warning system for disease indication can have its outputs appended to a blockchain, providing an auditable sequence of proof that can be tied to certificates. Friction-free collaboration between secure ledgers and analysis engines maintains the integrity of data from source to insight, promoting higher stakeholder trust levels. Cloud computing supports this integrated system, providing scalable storage, computing power, and network access, allowing for the support of decentralized, data-intensive aquaculture operations. Cloud architectures support the real-time ingestion of IoT sensor data, run high-end analytics applications for predictive calculations, and enable the deployment of blockchain nodes without the constraint of local infrastructure (Andrew et al., 2023).

Furthermore, cloud services also offer flexibility and affordability, allowing farms of all sizes to implement cutting-edge technologies without incurring heavy capital expenditures on IT infrastructure. The outcome of such a consolidation is guaranteed through real-time decision-making on an integrated digital platform. Operators and farmers can access analytics-driven dashboards in real-time, based on big data, and make informed decisions based on validated data streams. These decisions can then be recorded on the blockchain for future optimization and accountability. This closed-loop, dynamic system not only improves operations as more efficient and sustainable but also creates a new standard for transparency and resilience in global aquaculture practices (Gill et al., 2024).

While there is great potential for blockchain, cloud computing, and big data to be harnessed in aquaculture, numerous challenges must be addressed before these technologies can be utilized efficiently and sustainably. A significant technical barrier is interoperability, which refers to the ability of digital devices, platforms, and systems to communicate with each other and exchange information freely. In modern aquaculture environments, proprietary software, fragmented sensor networks, and disparate data formats often exist in silos, which hinder the flow of information and integration. Similarly, scalability issues are also faced as systems scale up; handling millions of real-time sensor points while maintaining performance and reliability is a significant bottleneck. Latency, especially in cloud-based architectures, can also interfere with time-critical operations, such as disease outbreak detection or automated feeding. Data privacy issues and regulatory barriers are additional challenges. As farms become increasingly dependent on digital platforms for sensitive operational information, protection against cyberattacks, data breaches, and unauthorized use becomes crucial. In addition, aquaculture tends to be trans-border in application across jurisdictions where data protection law, such as the GDPR within Europe, imposes high compliance thresholds. Finding a balance between innovation and legal and ethical pressures will be key to establishing trust and achieving large-scale adoption (Chandran et al., 2025). The need for standardization and international collaboration is increasingly urgent. Standardized protocols for data collection, processing, exchange, and safeguarding are needed to enable interoperability, ensure data quality, and facilitate multistakeholder collaboration along the aquaculture value chain. International cooperation among farmers, technologists, researchers, and policymakers will be necessary to establish these standards and promote best practices. In the future, prospects are on the rise. AI-powered blockchain analysis may offer intelligent, automated auditing of supply chains, enhancing both efficiency and transparency. Synergy between edge computing and the cloud, marrying on-site edge computing with central storage in the cloud, can achieve low latency and preserve computational intensity. Perhaps most intriguingly, decentralized autonomous aquafarms—self-managing, blockchain-enabled farm ecosystems—may emerge, operating with minimal human intervention and delivering maximum sustainability, resilience, and profitability (Akram et al., 2026).

References

Ahmed, F., Bijoy, M. H. I., Hemal, H. R., & Noori, S. R. H. (2024). Smart aquaculture analytics: enhancing shrimp farming in Bangladesh through real-time IoT monitoring and predictive machine learning analysis. *Heliyon, 10*(17), Article e37330. https://doi.org/10.1016/j.heliyon.2024.e37330

Akram, W., Din, M. U., Saad Saoud, L., & Hussain, I. (2026). A review of generative AI in aquaculture: applications, case studies and challenges for smart and sustainable farming. *Aquacultural Engineering, 112*. https://doi.org/10.1016/j.aquaeng.2025.102637

Arora, A., & Arora, A. (2022). Generative adversarial networks and synthetic patient data: current challenges and future perspectives. *Future Healthcare Journal, 9*(2), 190–193. https://doi.org/10.7861/fhj.2022-0013

Ashraf Rather, M., Ahmad, I., Shah, A., Ahmad Hajam, Y., Amin, A., Khursheed, S., … Rasool, S. (2024). Exploring opportunities of artificial intelligence in aquaculture to meet increasing food demand. *Food Chemistry X, 22*, Article 101309. https://doi.org/10.1016/j.fochx.2024.101309

Bunting, S. W., Bostock, J., Leschen, W., & Little, D. C. (2023). Evaluating the potential of innovations across aquaculture product value chains for poverty alleviation in Bangladesh and India. *Frontiers in Aquaculture, 2*. https://doi.org/10.3389/faquc.2023.1111266

Casino, F., Lopez-Iturri, P., & Patsakis, C. (2025). Cloud continuum testbeds and next-generation ICTs: trends, challenges, and perspectives. *Computer Science Review, 56*. https://doi.org/10.1016/j.cosrev.2024.100696

Chandran, P. J. I., Khalil, H. A., Hashir, P. K., & Veerasingam, S. (2025). Smart technologies in aquaculture: an integrated IoT, AI, and blockchain framework for sustainable growth. *Aquacultural Engineering, 111*. https://doi.org/10.1016/j.aquaeng.2025.102584

Chen, C. H., Wu, Y. C., Zhang, J. X., & Chen, Y. H. (2022). IoT-Based fish farm water quality monitoring system. *Sensors, 22*(17). https://doi.org/10.3390/s22176700

Corsi, A., de Souza, F. F., Pagani, R. N., & Kovaleski, J. L. (2021). Big data analytics as a tool for fighting pandemics: a systematic review of literature. *Journal of Ambient Intelligence and Humanized Computing, 12*(10), 9163–9180. https://doi.org/10.1007/s12652-020-02617-4

Cromwell, J., Turkson, C., Dora, M., & Yamoah, F. A. (2025). Digital technologies for traceability and transparency in the global fish supply chains: a systematic review and future directions. *Marine Policy, 178*. https://doi.org/10.1016/j.marpol.2025.106700

Getahun, S., Kefale, H., & Gelaye, Y. (2024). Application of precision agriculture technologies for sustainable crop production and environmental sustainability: a systematic review. *The Scientific World Journal, 2024*, Article 2126734. https://doi.org/10.1155/2024/2126734

Gill, S. S., Wu, H., Patros, P., Ottaviani, C., Arora, P., Pujol, V. C., … Buyya, R. (2024). Modern computing: vision and challenges. *Telematics and Informatics Reports, 13*. https://doi.org/10.1016/j.teler.2024.100116

Hegde, A., Kabra, S., Basawa, R. M., Khile, D. A., Abbu, R. U. F., Thomas, N. A., … Raval, R. (2023). Bacterial diseases in marine fish species: current trends and future prospects in disease management. *World Journal of Microbiology and Biotechnology, 39*(11), 317. https://doi.org/10.1007/s11274-023-03755-5

Hussein, J. B., Workneh, T. S., Kassim, A., Ntsowe, K., Melesse, S. F., & El-Mesery, H. S. (2025). A review on the impact of big data analytics in transforming agricultural practices, food processing, and preservation strategies. *Applied Food Research, 5*(2). https://doi.org/10.1016/j.afres.2025.101234

Ismail, S., Reza, H., Salameh, K., Kashani Zadeh, H., & Vasefi, F. (2023). Toward an intelligent blockchain IoT-Enabled fish supply chain: a review and conceptual framework. *Sensors, 23*(11). https://doi.org/10.3390/s23115136

Andrew, J., Isravel, D. P., Sagayam, K. M., Bhushan, B., Sei, Y., & Eunice, J. (2023). Blockchain for healthcare systems: architecture, security challenges, trends and future directions. *Journal of Network and Computer Applications, 215.* https://doi.org/10.1016/j.jnca.2023.103633

Khan, S. N., Loukil, F., Ghedira-Guegan, C., Benkhelifa, E., & Bani-Hani, A. (2021). Blockchain smart contracts: applications, challenges, and future trends. *Peer-to-Peer Networking and Applications, 14*(5), 2901–2925. https://doi.org/10.1007/s12083-021-01127-0

Luna, M., Fernandez-Vazquez, S., Tereñes Castelao, E., & Arias Fernández, Á. (2024). A blockchain-based approach to the challenges of EU's environmental policy compliance in aquaculture: from traceability to fraud prevention. *Marine Policy, 159.* https://doi.org/10.1016/j.marpol.2023.105892

Mohd Jais, N. A., Abdullah, A. F., Mohd Kassim, M. S., Abd Karim, M. M., Abdulsalam, M., & Muhadi, N. (2024). Improved accuracy in IoT-Based water quality monitoring for aquaculture tanks using low-cost sensors: Asian seabass fish farming. *Heliyon, 10*(8), Article e29022. https://doi.org/10.1016/j.heliyon.2024.e29022

Musa, H. S., Krichen, M., Altun, A. A., & Ammi, M. (2023). Survey on blockchain-based data storage security for android Mobile applications. *Sensors, 23*(21). https://doi.org/10.3390/s23218749

Neethirajan, S. (2025). Safeguarding digital livestock farming - a comprehensive cybersecurity roadmap for dairy and poultry industries. *Frontiers in Big Data, 8,* Article 1556157. https://doi.org/10.3389/fdata.2025.1556157

Nguyen, T., Nguyen, H., & Nguyen Gia, T. (2024). Exploring the integration of edge computing and blockchain IoT: Principles, architectures, security, and applications. *Journal of Network and Computer Applications, 226.* https://doi.org/10.1016/j.jnca.2024.103884

Pakseresht, A., Ahmadi Kaliji, S., & Hakelius, K. (2024). Blockchain technology characteristics essential for the agri-food sector: a systematic review. *Food Control, 165.* https://doi.org/10.1016/j.foodcont.2024.110661

Patel, A. S., Brahmbhatt, M. N., Bariya, A. R., Nayak, J. B., & Singh, V. K. (2023). Blockchain technology in food safety and traceability concern to livestock products." *Heliyon, 9*(6), Article e16526. https://doi.org/10.1016/j.heliyon.2023.e16526

Pradeepkiran, J. A. (2019). Aquaculture role in global food security with nutritional value: a review. *Translational Animal Science, 3*(2), 903–910. https://doi.org/10.1093/tas/txz012

Pranto, T. H., Noman, A. A., Mahmud, A., & Haque, A. B. (2021). Blockchain and smart contract for IoT enabled smart agriculture. *PeerJ Computer Science, 7,* Article e407. https://doi.org/10.7717/peerj-cs.407

Ramadan, M. N. A., Ali, M. A. H., Khoo, S. Y., Alkhedher, M., & Alherbawi, M. (2024). Real-time IoT-powered AI system for monitoring and forecasting of air pollution in industrial environment. *Ecotoxicology and Environmental Safety, 283.* https://doi.org/10.1016/j.ecoenv.2024.116856

Rejeb, A., Rejeb, K., & Keogh, J. G. (2025). The nexus of IoT and aquaculture: a bibliometric analysis. *Applied Food Research, 5*(1). https://doi.org/10.1016/j.afres.2025.100838

Rowan, N. J., Murray, N., Qiao, Y., O'Neill, E., Clifford, E., Barceló, D., & Power, D. M. (2022). Digital transformation of peatland eco-innovations ('Paludiculture'): Enabling a paradigm shift towards the real-time sustainable production of 'green-friendly' products and services. *Science of The Total Environment, 838*. https://doi.org/10.1016/j.scitotenv.2022.156328

Sedlakova, J., Daniore, P., Horn Wintsch, A., Wolf, M., Stanikic, M., Haag, C., … University of Zurich Digital Society Initiative Health, Community. (2023). Challenges and best practices for digital unstructured data enrichment in health research: a systematic narrative review. *PLOS Digital Health, 2*(10), Article e0000347. https://doi.org/10.1371/journal.pdig.0000347

Sivarajah, U., Kamal, M. M., Irani, Z., & Weerakkody, V. (2017). Critical analysis of big data challenges and analytical methods. *Journal of Business Research, 70*, 263–286. https://doi.org/10.1016/j.jbusres.2016.08.001

Thilakarathne, N. N., Abu Bakar, M. S., Abas, P. E., & Yassin, H. (2025). Internet of things enabled smart agriculture: current status, latest advancements, challenges and countermeasures. *Heliyon, 11*(3). https://doi.org/10.1016/j.heliyon.2025.e42136

Ullah, F., He, J., Zhu, N., Wajahat, A., Nazir, A., Qureshi, S., … Dev, S. (2024). Blockchain-enabled EHR access auditing: enhancing healthcare data security. *Heliyon, 10*(16), Article e34407. https://doi.org/10.1016/j.heliyon.2024.e34407

Wang, C., Li, Z., Wang, T., Xu, X., Zhang, X., & Li, D. (2021). Intelligent fish farm-the future of aquaculture. *Aquaculture International, 29*(6), 2681–2711. https://doi.org/10.1007/s10499-021-00773-8

Wangen, S. R., Zhang, F., Fadul-Pacheco, L., da Silva, T. E., & Cabrera, V. E. (2021). Improving farm decisions: the application of data engineering techniques to manage data streams from contemporary dairy operations. *Livestock Science, 250*. https://doi.org/10.1016/j.livsci.2021.104602

Zalik, K. R., & Zalik, M. (2023). A review of federated learning in agriculture. *Sensors, 23*(23). https://doi.org/10.3390/s23239566

13

Aquaculture and AI integration in industrial aspects

Introduction

Aquaculture involves breeding and harvesting aquatic organisms to meet global food demands. The aquaculture industry is facing numerous challenges because of the increasing global demand for seafood. The challenges include effective management of resources, optimization of manufacturing processes, disease management, and environmental sustainability. To solve these issues, the aquaculture sector started to use technological advancements, predominantly artificial intelligence (AI), to improve operational efficacy, cost reduction, and progress the overall output. The incorporation of AI into industrial aquaculture is renovating the landscape of fish farming and aquaculture organizations, donating pioneering solutions to multiple problems (Fig. 13.1). The embracing of AI in aquaculture encompasses the practice of machine learning (ML), deep learning (DL), neural networks, computer vision, data analytics, and robotics to enhance the multiple features of aquacultural operations. AI technologies are being used in the screening of water quality, forecasting fish growth patterns, systematizing feeding schedules, early disease detection, and certifying environmental situations. This incorporation offers to improve the policymaking process, enlarge manufacturing harvests, and maintainable practices that assist equally makers and patrons (Li et al., 2021). In addition, AI-driven systems investigate enormous amounts of data collected from sensors, satellite images, and monitoring devices, as well as visions that would otherwise be unachievable through traditional methods. One of the most important aids of AI to industrial aquaculture is in the field of fish health management. Outmoded disease-finding methods in aquaculture habitually depend on manual observation, which is inefficient and susceptible to human error. With the combination of AI, though, initial disease detection turns out to be more specific and effective. AI algorithms evaluate patterns in fish performance, environmental

Aquaculture with Artificial Intelligence. https://doi.org/10.1016/B978-0-443-33704-8.00003-3

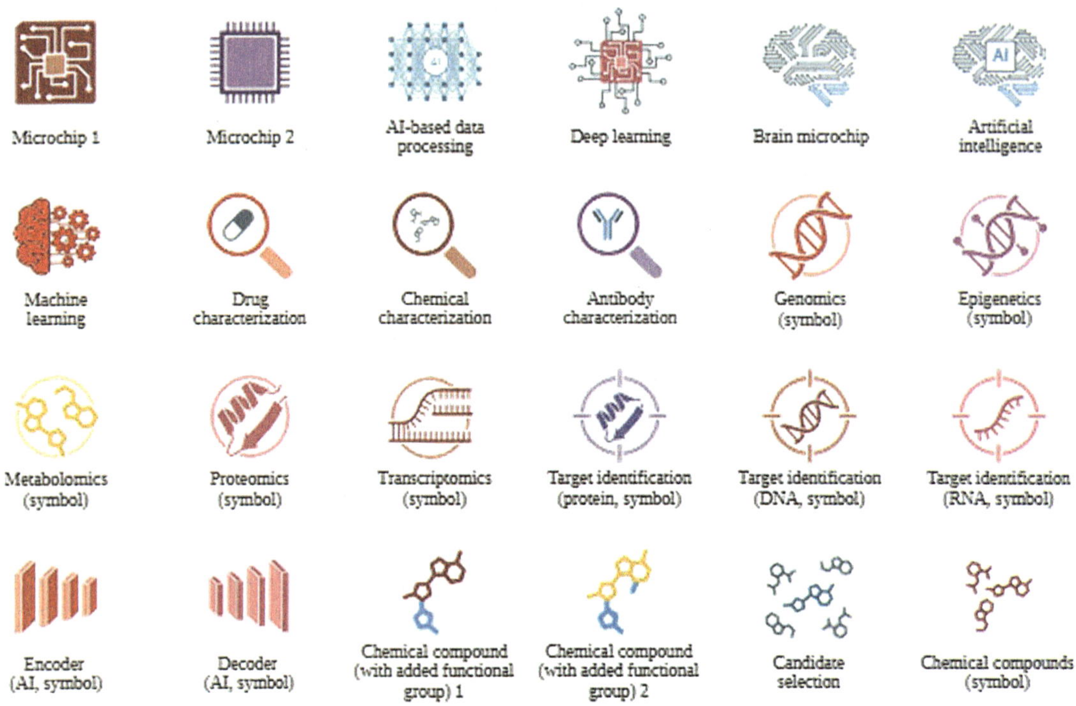

Figure 13.1 Various technological advancements in the form of AI for industrial aspects.

circumstances, and other variables to forecast possible disease epidemics before they break out. For instance, AI-motorized computer revelation systems screen fish movements and categorize abnormalities in their behavior, including compact action or irregular swimming, which may point to the existence of a disease (Zhou et al., 2022). In addition, AI systems are used to comment on proper instructions, for instance, water quality parameter alteration or ordering of best treatments, further minimizing the outbreak of diseases. AI plays a role in feeding practices in another critical field of concentration. Feeding is a substantial cost factor in aquaculture, secretarial for a considerable share of the operational expenses. Overfeeding or underfeeding can cause growth deficiency, excess leftovers, and environmental pollution. AI-powered feeding systems examine the real-time data on water quality and fish behavioral patterns to regulate the finest volume of feed offered at any given time. This data determines methods to aid in the reduction of feed wastage, fish growth improvement, and enhance feed conversion ratios (Zhang et al., 2020a, 2020b). Moreover, an AI-powered system adjusts feeding plans based on the fish growth stages, certifying that the feeding procedure is improved for overall productivity. Furthermore, in health and feeding, AI also plays a major role in environmental management and monitoring. The water quality in the aquaculture system is vital to the health and growth of aquatic organisms. Consequently, parameters such as temperature, salinity, oxygen levels, and pH must be regularly monitored to confirm they remain within the optimal ranges. Old methods for water quality monitoring are labor-dependent and do not exhibit real-time data, which is vital for decision-making at the exact time. AI offers continuous real-time monitoring of the water quality and adjusts the system's parameters promptly using sensors and the Internet of Things (IoT). AI-enabled real-time monitoring of water quality not only improves the well-being of the aquatic organisms, but it also contributes to environmental sustainability by waste reduction and minimization of waste production of aquaculture systems, which do not affect the surrounding ecosystems (Dai et al., 2021). The ability for AI to transform industrial aquaculture spreads to computerization in aquaculture amenities. Robotics and automated systems are gradually being combined into aquaculture, performing duties such as reaping, sorting, and cleaning. This system not only reduces the labor expenses but also improves the accuracy and competence of operations. Perhaps, autonomous underwater vehicles (AUVs) armed with AI are used for examining fish cages, assessing the health of the fish, and even doing treatments if essential. These robotic systems work in exacting underwater environments, where human activities may be hard or dangerous, thus escalating the series of potential risks for aquaculture actions (Liu et al., 2022). Though the addition of AI in aquaculture offers abundant aids, it also elevates concerns about data security, privacy, and the need for particular assistance. The high quantity of data created through AI systems entails a strong setup for storage, processing, and assessment. Besides, there are worries around the possibility of data holes and the misuse of subtle information, mainly in relation to competitive intelligence. As well, the acceptance of AI technologies in aquaculture needs skilled professionals who create, maintain, and interpret AI-powered systems. Consequently, there is an emerging requirement for educational programs and working on AI applications within the aquaculture sector to confirm that the workforce is equipped with the essential skills (Harper et al., 2021).

Role of artificial intelligence in modern aquaculture

The incorporation of AI into the current aquaculture sector has proliferated in a reformative era in the organization and streamlining of aquatic farming sectors. Aquaculture, the rearing of marine animals, including fish, shellfish, and algae, has seen a lot of challenges associated with environmental sustainability, reserve administration, and working productivity. AI, done, it is data-powered and progressive learning algorithms, plays a vital role in rectifying these challenges, resulting in considerable improvements in manufacturing productivity, resource management, and sustainability within the sector. AI technologies, mainly ML and DL, have reformed the key view of aquaculture. One of the most effective deployments in fish health management. AI-driven systems are skilled at assessing large amounts of data from environmental sensors, video feeds, and other monitoring devices to sense early signs of disease epidemics, irregular characteristics, and stress in aquatic animals. AI assesses the health and growth of the animals in real time by implementing the fastest interventions and decreasing reliance on chemicals or antibiotics. This practical method of disease management not only improves the health of the aquaculture stock but also decreases environmental pollution due to the high-dose chemical usage. Added sector AI has been crafted significant strides in feeding systems. AI algorithms are progressively being utilized to organize feed dispersal based on real-time data on water quality, behavior of fish, and growth. Outmoded aquaculture feeding frequently results in overfeeding or underfeeding. This is resulting in inefficiencies and environmental pollution. AI-powered feeding system permits particular feeding, decreases the feed waste, intensifies the fish growth rates, and abates the ecological footprint of the process. Endlessly learning from historical data and regulating feeding applies, these systems help attain an equilibrium between output and sustainability. AI also helps in environmental assessment and optimization in aquaculture farms. Water quality, predominantly constraints as well as temperature, pH, dissolved oxygen levels, and ammonia absorption, is vital for the welfare of aquatic fauna. AI-ambitious prognostic models and instantaneous analytics are cast off to investigate these constraints more indeed and to forecast potential difficulties before they turn out to be perilous. Through merging IoT radars with AI algorithms, aquaculture machinists keep the finest ecological circumstances, emerging total manufacturing, and reducing the hazard of system disappointments that result in fish death or lesser yields. In terms of systematization and robotics, AI-motorized systems are employed to automate errands such as sorting, grading, and harvesting. These responsibilities are habitually essential workforce activities, and employing AI-permitted machines results in improved meticulousness and competence. Besides, AI-powered systems are cast off to calculate the finest harvesting time based on fish development rates and market need, confirming that the harvest process is both economically and ecologically ideal. AI plays a role in improving supply chains, which is a further decisive advantage. AI empowers aquaculture rearing units to exactly track production, manage inventories, and forecast market trends. Using AI to augment supply chains rallies decision-making in production planning, logistics, and marketing, boosting profitability and competitiveness. In addition, AI supports sustainable practices by helping farms minimize ecological impacts. Data analytics and AI insights allow farms to optimize water use, reduce waste, and better manage energy resources. Combining AI with sustainability efforts is crucial for transitioning to eco-friendly, resource-efficient aquaculture methods, which are vital for meeting rising global seafood demand while protecting marine ecosystems.

AI in water quality monitoring and management

The incorporation of AI in water quality monitoring and management has reformed the technique of aquatic ecosystems, predominantly in aquaculture, which is monitored, controlled, and augmented (Fig. 13.2). Water quality is vital for the viability of

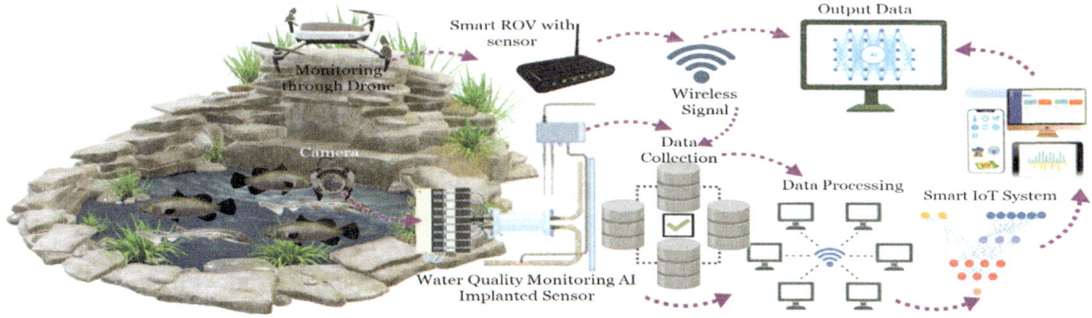

Figure 13.2 Integration of artificial intelligence (AI) in water quality monitoring and management.

aquacultural applications, as it directly affects the aquatic animals' health and the productivity of the aquacultural maneuvers. Old-style approaches of monitoring water quality, which depend on manpower, including sampling and laboratory examination, are laborious and prone to delays, making them insufficient for real-time, constant monitoring. AI-powered results report these confines by providing intelligent systems that assess the water quality parameters, such as temperature, pH, dissolved oxygen, turbidity, salinity, and ammonia levels, which are crucial for aquaculture procedures. AI machinery, including ML, DL, and data analytics, plays a major role in the on-time forecast and valuation of water quality circumstances. These systems examine massive amounts of environmentally related data collected through sensor arrangements arrayed across aquaculture substructures, permitting precise and opportune forecasts of water quality variations and variance detection. As AI-ambitious models anticipate potential harmful algal blooms (HABs), changes in dissolved oxygen concentrations, or sudden variations in pH levels, they enable anticipatory interferences. ML techniques, predominantly those using controlled learning, are applied to antique datasets to recognize patterns and relationships between environmental conflicts and aquatic animal well-being, thus enhancing the meticulousness of water quality oversight. To further improve irregularity detection and prognostication, AI rationalizes water treatment procedures via mechanization, vigorously regulating variables including aeration and filtration in response to simultaneous data contributions. This decreases dependence on labor-intensive oversight and increases working productivity. Furthermore, AI empowers seamless incorporation of remote detection and IoT technologies for incessant aquatic screening. IoT-allowed sensor devices, deliberately situated within aquaculture ranches, collect various water quality metrics and communicate them to AI appliances for instantaneous analysis, distributing actionable insights through spontaneous dashboards or mobile platforms. The potential to aggregate and construe multisource data fosters a more all-inclusive understanding of aquatic system subtleties, decontaminating policymaking and permitting more groundbreaking management approaches. Also, AI aids the adoption of meticulous aquaculture, tailoring water quality control to the exact biological needs of different aquatic organisms, improving environmental well-being, and enhancing productivity. In conclusion, AI-driven water management backgrounds sustainability by dropping resource utilization and minimalizing environmental footprints. Perhaps, AI augments the practice of chemicals and other medications, plummeting the multiple uses of pesticides or fertilizers that destroy the aquatic atmosphere. It also improves the capability to preserve optimal environmental surroundings, which provisions the long-term viability of aquaculture sectors and prevents ecological deprivations.

Automation and robotics in aquaculture operations

Automation and robotics are progressively transmuting the aquaculture sector, improving both operational efficacy and longevity. Aquaculture, which encompasses the rearing of aquatic creatures, including fish, shellfish, and algae, faces multiple challenges such as insufficient manpower, environmental impacts, and the requirement for competent resource management. Automation and robotics discourse these encounters by improving multiple stages of aquaculture procedures, from nourishing and nursing to garnering and dispensation. One of the most substantial applications of automation in aquaculture is in the feeding process. Automatic alimentation systems employ sensors, algorithms, and ML to confirm that fish obtain the optimum quantity of food at the accurate periods, thus reducing waste and improving growth rates. These schemes screen water circumstances and fish behavior, regulating the feeding agendas based on instantaneous data. The combination of AI permits these schemes to acquire and acclimate to fluctuations in fish behavior and conservational circumstances, subsequent in further particular alimentation maneuvers (Jiang et al., 2021). Robotic systems have also transformed one-to-one care and scrutiny in aquaculture. AUVs and drones are extensively utilized for simultaneous intensive care of water superiority, fish health, and the veracity of ranch substructure. These robots are armed with cameras, sensors, and further analytical tools that distinguish premature symptoms of disease, algal blooms, or added damaging environmental aspects, aiding agronomists to retort proactively. AUVs correspondingly amass samples of water and sediment, providing treasured data for scientific study and grange management (Olivar et al., 2022). The usage of robotics in aquaculture is inadequate for one-to-one care and alimentation; nonetheless, outspreads to garnering and dispensation. Robotic reapers decrease the labor-intensive procedure of catching and handling fish. These systems employ accurate algorithms and automatic arms to lightly hold the fish, abating stress and damage. In addition, robotics automates postharvest handling, together with grading, sorting, and packaging, guaranteeing reliability in merchandize value and reducing human fault (Rodríguez et al., 2020). The incorporation of automation and robotics in aquaculture gives abundant paybacks. These technologies upsurge manufacturing efficacy by diminishing human interference, plummeting operational budgets, and refining complete merchandize eminence. They also subsidize environmental longevity by dropping leftover, improving resource usage, and permitting added sustainable agriculture performs. For instance, accurate feeding decreases the quantity of uneaten food in the water, which in shot depresses nutrient contamination and encourages improved water quality (Foster et al., 2021). In spite of the paybacks, the implementation of automation and robotics in aquaculture raises some concerns. A lot of early venture expenses and the necessity for specific familiarity to preserve and activate progressive systems are substantial blockades for minor ranches. Besides, apprehensions occur near the conservational effect of extensive mechanization, as well as energy ingesting and the probable for system breakdowns. However, progressions in AI and ML are construction these systems more profitable and unswerving, flooring the way for widespread implementation (Jiang et al., 2021).

Disease prediction and health management through AI

Aquaculture, a dynamic sector of worldwide food production, airs plentiful encounters, predominantly in handling fish health and preventing disease epidemics. In recent years, AI has arisen as a prevailing device in improving sickness forecast and healthcare in aquaculture organizations. AI machinery, including ML, DL, and computer vision (CV), has demonstrated influence in the initial recognition and supervision of disease, which are habitually a substantial hazard to aquaculture manufacturing. This machinery aids instantaneous intensive care and investigation of innumerable well-being constraints in aquaculture ranches, serving to avert the blowout of diseases, decrease victims, and progress complete ranch efficiency. AI-powered systems evaluate massive amounts of data engendered by sensors, cameras, and ecological monitoring tools, provided that intuitions into water quality, temperature, oxygen levels, and the behavior of aquatic creatures. For example, computer vision technologies have been used to monitor the physical presence and behavior of fish, sensing elusive variations such as irregular swimming patterns, discoloration, or scratches that may indicate the beginning of illnesses (Li et al., 2020). ML models then examine these pictorial signs and ecological data to forecast the prospect of illness epidemics, clinically ostensible signs. Initial discovery of AI permits timely interpositions, including regulating ecological circumstances or administering beset treatments, which significantly decreases the necessity for broad-spectrum antibiotics and chemicals. Besides, AI has empowered prognostic modeling systems, which predict the spread of disease inside aquaculture systems by mixing ancient data on illness incidence, water conditions, and fish species behavior. These analytical models are cast off to evaluate the danger related to latent epidemics and augment resource distribution for disease surveillance (Rodríguez et al., 2019). For instance, AI-driven algorithms predict possible hazards connected to environmental factors, including water temperature variations, which are known to stimulate the commonness of precise pathogens such as *Vibrio* and *Aeromonas* species, in aquaculture systems. Forecasting these menaces, ranchers take practical procedures to alleviate the impact of illness epidemics. AI's advantages in well-being supervision encompass their disease prediction; it also comprises augmenting the general well-being and development of aquaculture organisms. Through AI, meticulousness feeding systems are applied, where data from sensors are assessed to deliver customized diet plans for fish, confirming that apiece fish obtains the proper quantity of forage based on its growing stage and well-being position. As long as exactness feeding systems not only endorse healthier fish but also improve the efficacy of nourishing performance, decreasing leftover and depressing expenses (Zhang et al., 2021). The combination of AI into aquaculture healthiness supervision systems is also improving longevity in manufacturing. Plummeting dependence on chemical handlings and antibiotics, AI helps minimize ecological effluence and avert the expansion of antibiotic resistance. The instantaneous monitoring competencies provided by AI also permit healthier executives concerning fish keeping compactness, water quality management, thus augmenting output while preserving the well-being of both aquatic organisms and the environment.

Data analytics and decision support systems

The incorporation of data analytics and decision support systems (DSS) in aquaculture has appeared by way of an essential progression, lashing the manufacturing in the production toward superior competence and longevity. Data analytics in aquaculture incorporates the congregation, processing, and examination of enormous extents of data from various sources,

Including environmental radars, water quality monitors, and feed alimentation patterns. Large data procedures, ML, and AI permit aquaculture procedures to make acquainted findings that enhance fish growth, improve feeding capability, and enhance illness supervision. Rapid data dealing out permits the monitoring of variable measures, such as water temperature, salinity, pH levels, dissolved oxygen, nutrient levels, and delivers a comprehensive interpretation of the aquatic environment. These rapid perceptions empower operatives to regulate effective restrictions proactively, thus minimizing danger and exploiting yield. DSS in aquaculture performance a perilous role in the operative employment of data analytics by providing decision-makers with the utensils to construct and act on data. DSS syndicates data, models, and algorithms to excite numerous situations and forecast consequences under dissimilar circumstances. In the background of aquaculture, DSS is consumed to improve feeding schedules, forecast fish growth rates, choose the optimum reaping time, and augment inclusive ranch supervision. Furthermore, by mixing ML and AI techniques, DSS can forecast illness epidemics by evaluating learnings in ecological circumstances, water quality, and ancient well-being data. These prognostic competences permit ranchers to gross problems previously, preventing them from worsening, plummeting the number of sufferers, and guaranteeing the longevity of fish farming maneuvers. The role of AI in these systems cannot be exaggerated. AI-powered models, including neural networks and DL algorithms, interpret huge datasets much quicker and with superior accuracy than outmoded approaches. These models also recognize concealed outlines in data that would then go ignored, opening new prospects for optimization. Perhaps, AI envisages the growth patterns of numerous fish species grounded on ecological variable quantity, regulates alimentation rates in real-time to decrease leftovers, and estimates the probability of disease epidemics grounded on ecological circumstances and ancient data. As well, the incorporation of AI with IoT sensors allows for constant nursing and computerization of aquaculture maneuvers, making them more accurate and less dependent on human interference. The interaction among data analytics, DSS, and AI in aquaculture is altering the trade landscape by improving yield, decreasing operational expenses, and enlightening ecological longevity. As AI technology endures to change, its amalgamation into aquaculture will be expected to develop progressively urban, permitting ranchers to make more conversant, data-powered verdicts that warrant the long-term achievement and cost-effectiveness of aquacultural maneuvers.

Supply chain optimization and market forecasting

Supply chain optimization and market forecasting in aquaculture are essential features that encourage the productivity, permanency, and efficiency of the manufacturing sector. Aquaculture, as a growing and crucial area for global nutrition, faces significant challenges in ensuring optimal nutrient consumption and maintaining the integrity of its supply chain. Integrating AI into these processes offers revolutionary potential. AI-motorized equipment examines large quantities of data from various sources, including environmental conditions, production rates, logistics, and market demand, to forecast trends, improve operations, and inform more knowledgeable management decisions. Improving the aquaculture supply chain comprises reorganizing manufacturing and circulation systems to decrease leftovers, lessening charges, and improving product quality. AI supports these laborers by using progressive algorithms and ML models to forecast market tendencies and augment the register controller. As AI can investigate ancient data on fish manufacturing, customer demand, and logistics to endorse effective delivery timetables and ways, thus minimizing postponements and degradation. Moreover, AI utensils screen key parameters including water quality, feed conversion proficiency, and stock well-being, permitting more precise prognostication and improved resource distribution. Market prognostication vital for both manufacturers and providers has habitually depended on professional decision and ancient patterns, but such approaches frequently lack exactness and adaptability to altering circumstances. In contrast, AI-ambitious prediction influences huge datasets, counting climate trends, consumer behavior, and socioeconomic indicators, to distribute more dependable forecasts. In aquaculture, these potential permits manufacturers to anticipate mandate changes for fish, shellfish, and other products, altering manufacturing proactively. Precise prognostication is predominantly dynamic due to the delicate nature of aquaculture yields, where overproduction leads to waste and underproduction results in misplaced sales prospects. AI-driven market forecasts also support supply chain resilience by incorporating data on peripheral effects, including geopolitical problems, trade procedures, and environmental disturbances. This dynamic method benefits aquaculture businesses to respond more effectually to hazards and worries. Furthermore, AI systems can forecast customer preferences and acquiring habits, permitting manufacturers to adapt to developing trends like the demand for longevity farmed seafood or plant-based replacements. The application of AI in aquaculture not only fosters sustainability but also progresses economic feasibility. For instance, AI-based feed optimization averts overfeeding, depressing environmental impacts, while advanced market prognostication decreases inadequacies, including stock shortages or surpluses.

Sustainability and environmental impact mitigation

Sustainability and environmental impact temperance are decisive trepidations in the context of present aquaculture, predominantly when combining AI into industrial practices. The hasty development of international aquaculture to encounter cumulative seafood demand has elevated worries about the manufacturing's continuing feasibility, primarily concerning resource utilization, environmental degradation, and its impacts on biodiversity. Outdated aquaculture habitually results in overfishing, water contamination, habitat devastation, and illness epidemics, subsequent in substantial environmental consequences. These problems are exacerbated by the practice of exhaustive farming performs and dependence on wild-caught fish for feed, further weakening marine environments. AI technologies provide capable solutions to rectify these environmental complications by enhancing resource utilization and decreasing leftover. For instance, AI-driven systems screen water quality key parameters, including oxygen, temperature, and pH levels simultaneously, empowering more operative management of water circumstances. Primary estimate of water quality problems with AI aids prevents complications corresponding to hypoxia or acidification, which portend to the marine ecosystem and can lead to extensive fish death rates. Moreover, AI augments feed supervision by precisely forecasting the dietary requirements of cultivated aquatic organisms, thus decreasing the amount of extra feedstuff that contaminates water. Also, AI-ambitious systems expand sickness discovery and management, enabling initial documentation of potential epidemics and plummeting reliance on antibiotics, which can destroy the environment and human well-being. In addition, AI can subordinate the carbon footprint of aquaculture maneuvers. Via ML algorithms and big data analytics, AI improves energy utilization, evaluates ecological effects, and provides insights for plummeting the inclusive environmental footprint of ranches. For example, AI-based predictive models support the expansion of justifiable setups such as recirculating aquaculture systems (RAS), which minimize water use and progress competence. These systems are more sustainable than traditional open-water farms because they reduce water waste, contamination, and the risk of invasive species escaping into the wild. Besides improving farm processes, AI also enhances supply chain management by improving logistics, reducing waste during transportation, and increasing traceability, helping aquaculture businesses operate more sustainably and with less environmental impact. Waste reduction is especially important in international markets where consumers demand sustainable seafood. AI-powered tools that assist with sustainability certifications, such as the Aquaculture Stewardship Council (ASC) and the Marine Stewardship Council (MSC), help ensure aquaculture products meet environmental standards and match consumer preferences for responsibly sourced seafood. Despite these advantages, integrating AI into aquaculture presents challenges to achieving full sustainability. A key concern is the accessibility of technology, as smaller farms may find AI systems too expensive, limiting their ability to operate efficiently. In addition, the rapid growth of AI raises concerns about unintended effects and overreliance on automation, which could reduce the nuanced understanding of environmental systems that human expertise offers. Therefore regulatory frameworks need to develop alongside technological advances to ensure AI use aligns with broader sustainability goals and is fair across the industry.

Challenges, opportunities, and future prospects

The integration of AI into aquaculture has created new opportunities to boost output, efficiency, and lifespan in the industry. However, it also brings significant challenges. One major difficulty is the high initial cost of AI technologies. Implementing AI systems, such as automated feeders, underwater drones, and AI sensors, often requires a large financial investment, which can be a barrier for small to medium-sized farms. The aquaculture industry is also very diverse, covering many species, environmental conditions, and operational scales. Developing AI systems that are flexible and adaptable across different setups remains tough. For instance, AI tools designed for fish farming in tropical climates may not work well in colder waters because of differences in species behavior and environmental factors. Data management is another key issue. AI systems depend heavily on large datasets for analysis and decision-making, but in aquaculture, data are often limited, noisy, or inconsistent. Gathering accurate, high-quality data from underwater sensors or cameras in real-time can be hard due to water turbidity, sensor failures, and environmental conditions. In addition, the complex interactions among factors such as water quality, fish health, and feed efficiency make AI models complicated, making it hard to develop predictive systems that handle the multidimensional nature of aquaculture data. Another challenge is the shortage of trained personnel. Successful AI adoption needs a workforce skilled in both aquaculture and AI technology. The skills gap slows down the wider use of AI innovations. Despite these obstacles, the benefits AI offers to aquaculture are considerable. AI-driven systems can greatly improve operational efficiency, automate routine tasks, and reduce labor costs. For example, AI sensors can monitor water quality in real-time, helping optimize conditions for fish health and growth. AI-based predictive models can also improve feeding schedules by analyzing data on fish behavior, water temperature, and other factors. This leads to less feed waste and more efficient use of resources, saving costs and benefiting the environment. AI systems also help detect and manage diseases by identifying early signs of outbreaks, which reduces the need for antibiotics and chemicals. Moving toward precision aquaculture is crucial for addressing concerns about the environmental impact of traditional methods. Looking to the future, AI's potential in aquaculture is promising. As technology progresses, AI systems will become more advanced and capable of handling complex, real-time decision-making. The integration of ML, DL, and other advanced AI techniques will improve accuracy and automation. Future developments might also include connections with new technologies, such as the IoT and blockchain. For example, IoT devices could supply continuous, real-time data to AI systems, boosting their decision-making power. Blockchain could improve transparency and traceability in the supply chain, encouraging sustainable and ethical practices in aquaculture. Furthermore, AI could play a key role in addressing global food security by increasing aquaculture productivity and efficiency. By optimizing resource management and farming practices, AI helps lessen pressure on wild fisheries and supports sustainable seafood production.

Conclusion

The incorporation of AI in aquaculture has led to significant progress in the industry, enhancing yield, sustainability, and resource management. This chapter emphasizes how AI technologies, including ML, data analytics, and automation, have transformed aquaculture by improving fish health monitoring, optimizing dietary plans, and ensuring efficient resource use. AI-powered systems enable real-time data analysis, allowing for precise decision-making in water quality management, disease detection, and predicting stock growth. In addition, AI-based models support the development of sustainable aquaculture systems by reducing environmental impacts and minimizing waste. As the industry evolves, AI implementation is expected to increase production capabilities, decrease operational costs, and promote ecological responsibility. However, realizing AI's full potential in aquaculture requires further research, innovation, and collaboration among academia, industry, and policymakers to address challenges related to scalability, data security, and ethical considerations.

References

Foster, C., Liu, X., & Zhang, Z. (2021). Advances in automated feeding systems for aquaculture. *Aquaculture Engineering, 94*, 1–10.

Harper, D., McAlpine, K., & Bower, M. (2021). Integrating artificial intelligence in aquaculture operations: a review of the current state and future opportunities. *Aquaculture Engineering, 88*, Article 102112.

Jiang, H., Wang, J., & Zhang, L. (2021). Intelligent feeding systems in aquaculture: a review. *Aquaculture Automation & Robotics, 45*(3), 21–29.

Li, Y., et al. (2020). Fish disease detection and prediction using machine learning and computer vision. *Aquaculture Engineering, 89*, 23–33.

Li, X., Zhang, H., & Wang, H. (2021). Artificial intelligence in aquaculture: a review. *Aquaculture, 533*, Article 736046.

Liu, H., Ma, Z., & Zhang, Q. (2022). The role of autonomous robots in modern aquaculture. *Journal of Robotics and Automation, 13*(4), 189–203.

Olivar, M. P., Gallego, V., & Martínez-Pérez, J. (2022). Robotics in aquaculture: underwater vehicles and drones for monitoring aquatic ecosystems. *Robotics and Automation in Aquatic Environments, 33*(4), 120–134.

Rodríguez, M., et al. (2019). Predictive modeling for disease outbreaks in aquaculture: a review. *Aquaculture Research, 50*(6), 1670–1684.

Rodríguez, R., Medina, M., & López, M. (2020). Robotics in fish harvesting: design and implementation of automated systems. *Journal of Aquatic Robotics, 12*(2), 45–59.

Dai, Y. S., Liu, S., & Chen, Z. (2021). Smart aquaculture: the role of artificial intelligence in water quality management. *Aquaculture Research, 52*(6), 2347–2358.

Zhang, Y., et al. (2021). AI-enabled precision feeding systems in aquaculture: a review. *Aquaculture Automation & Control, 13,* 100–112.

Zhang, H., Li, B., & Wang, F. (2020a). AI-based precision aquaculture: a comprehensive review. *Aquacultural Engineering, 88,* Article 102033.

Zhang, Y., Yang, L., & Sun, T. (2020b). AI-driven feeding management systems in aquaculture: enhancements and future prospects. *Aquaculture Engineering, 90,* Article 102112.

Zhou, Z., Wang, L., & Li, C. (2022). AI-based disease detection in aquaculture systems: a case study. *Aquaculture Systems, 18*(3), 111–121.

14

Future directions

Introduction to AI in aquaculture

Aquaculture, the rearing of aquatic creatures such as fish, mollusks, crustaceans, and aquatic plants, has been raised as a vital industry in rectifying worldwide food security concerns. As the world's population continues to increase, the need for longevity seafood manufacturing increases, demanding new resolutions to improve yield, confirm environmental sustainability, and progress financial feasibility. In current years, artificial intelligence (AI) has been getting significant attention as a revolutionary technology proficient of rectifying multiple encounters faced by the aquaculture industry. AI technologies, which include machine learning (ML), deep learning (DL), computer vision, robotics, and data analytics, are being extensively incorporated into aquaculture to augment feeding commands, supervise fish health, forecast illness epidemics, manage ecological circumstances, and machinery farming (Føre et al., 2018; Misimi et al., 2020). The use of AI in aquaculture is necessarily changing outdated agrobusiness practices by empowering exactness aquaculture, an archetype that highlights data-driven decision-making. Meticulousness in aquaculture depends on constant intensive care via sensors, submerged cameras, drones, and remote sensing machineries to gather massive quantities of data on water quality constraints, fish behavior, development rates, and illness signs (Zhang et al., 2022). AI algorithms handle and examine these datasets instantaneously to originate executable perceptions, allowing ranchers to make conversant verdicts that progress operative efficacy and decrease ecological influences. For example, ML models forecast optimum feeding periods and measures based on fish craving and behavior examination, minimizing food leftover and improving growth performance (Rokhlin et al., 2021). Congruently, computer visualization systems perceive early signs of disease, physical irregularities, or anxiety in fish populations, enabling appropriate interventions that pointedly decrease death rates. One of the most capable sectors of AI application in aquaculture is disease

Aquaculture with Artificial Intelligence. https://doi.org/10.1016/B978-0-443-33704-8.00007-0

finding and management. Conventionally, disease identification depends heavily on human examination and laboratory examination, progressions that are long, laborious, and habitually susceptible to interruptions. AI-driven analytical tools, particularly those using convolutional neural networks (CNNs), hastily examine images and video material of fish to perceive illness signs with high precision (Li et al., 2020). Additionally, prognostic analytics models, consuming ancient and conservational data, are being established to predict illness epidemics before they ensue, permitting proactive health management approaches. Such progressions not only improve the well-being and safety of cultured organisms but also subsidize to plummeting the necessity for antibiotics and chemicals, encouraging more supportable and environmentally friendly farming practices. Environmental monitoring and management are added perilous provinces where AI is demonstrating to be irreplaceable. Sustaining the finest water quality is vital for the accomplishment of any aquaculture maneuver; though, outmoded monitoring approaches frequently deliver only sporadic and inadequate data. AI systems armed with Internet of Things (IoT) strategies and independent daisies empower incessant, instantaneous monitoring of crucial water parameters, including temperature, dissolved oxygen, pH, salinity, and ammonia levels (Rajesh et al., 2021). By way of extrapolative modeling, AI predictions of vicissitudes in environmental circumstances warn ranchers of probable hazards and endorse counteractive actions. Furthermore, AI-powered independent vehicles, as well as submerged drones and robots, are being arrayed for responsibilities including net examination, biofouling elimination, and stock valuation, thus plummeting labor charges and refining ranch management performs. In spite of its revolutionary probable, the incorporation of AI in aquaculture is still in its emerging stages, in contrast to further manufacturing, including agriculture or manufacturing. Numerous encounters must be addressed to fully apprehend AI's aids in aquaculture, such as problems associated with data quality and availability, the high cost of applying AI systems, a deficiency of procedural proficiency amid ranchers, and apprehensions concerning data privacy and cybersecurity. In addition, the multiplicity of aquaculture organisms, agricultural systems, and geographical circumstances confounds the expansion of worldwide AI resolutions (Dawson et al., 2020). Hence, upcoming investigations should concentrate on emerging flexible, climbable, and profitable AI applications personalized to the precise requirements of numerous aquaculture industries. Setting sights on the forthcoming of AI in aquaculture seems auspicious, ambitious, with quick progressions in computational power, radar machinery, and large data analytics. Developing inclinations, including the incorporation of AI with blockchain for traceability, the use of digital twins for ranch replication and management, and the expansion of reasonable AI models to augment transparency and conviction among operators, are expected to shape the next generation of smart aquaculture systems (Xu et al., 2023). Furthermore, interdisciplinary partnership among computer scientists, aquaculture specialists, engineers, and policymakers will be vital in providing novelty, confirming the accountable disposition of AI, and exploiting its societal and environmental aids. Connecting the influence of AI, the aquaculture trade moves toward a more sustainable, buoyant, and productive future, contributing pointedly to worldwide food security and the blue economy.

Trends and technological advancements in AI

Current years have observed extraordinary drifts and technological progressions in AI that are premeditated to transform aquaculture systems. Key expansions comprise DL, computer vision, IoT incorporation, and edge computing, all of which are permitting more accurate, well-organized, and longevity aquaculture maneuvers. DL models, including CNNs, have been gradually espoused for instantaneous fish species sorting, behavioral monitoring, and illness discovery, thus minimalizing humanoid mistake and improving manufacturing upshots (Zhao et al., 2021). AI-driven computer vision technologies, combined with submerged cameras and drones, now enable constant health evaluation, biomass assessment, and automatic feeding systems founded on instantaneous examination of fish behavior and famine (Qin et al., 2020). Likewise, the incorporation of AI with IoT devices permits smart aquaculture ranches, where radars monitor key water quality parameters (e.g., pH, dissolved oxygen, and temperature), and AI algorithms forecast environmental oscillations, thus improving farm management (Neto et al., 2021). Additional auspicious leaning is the increase of advantageous AI, where data processing happens at the device level, tumbling expectancy, and guaranteeing sooner policymaking critical for active aquaculture environments (Shi et al., 2019). Strengthening learning algorithms is being explored to augment feeding agendas and energy intake in aquaculture operations, conducive to both economic and environmental longevity. Moreover, AI models are progressively integrating big data analytics from genomics, phenomics, and ecological sources to support selective breeding programs aimed at improving growth rates, disease resistance, and stress lenience in cultured species (Houston et al., 2020). For instance, AI technologies remain in development, ethical AI practices, explainable AI (XAI), and robust data governance are converting fundamental trends to confirm transparency, hope, and sanctuary in aquaculture inventions. These technical routes point toward a future where AI will not only augment operational efficacy but also play a crucial role in attaining longevity aquaculture that allies with international food security goals. Constant interdisciplinary partnership, vigorous governing frameworks, and speculation in AI learning among aquaculture consultants will be vital to completely connect the potential of these developing technologies.

Precision aquaculture: The next evolution

Precision aquaculture signifies the next momentous development in longevity fish farming, harnessing progressive technologies, predominantly AI, to augment production, improve animal well-being, and minimize environmental effect. This developing perception is based on the practicalities of outdated aquaculture, presenting a data-intensive, extremely restrained method that empowers ranchers to make instantaneous, informed conclusions. Participating AI algorithms with radar technologies, ML models, computer vision, and the IoT, meticulous aquaculture empowers the monitoring of a wide range of constraints, such as water quality, fish behavior, feeding patterns, and disease epidemics, with unparalleled precision (Føre et al., 2018). The outcome is a move away from widespread farming practices toward adapted, site-specific management systems that confirm the well-being and efficiency of aquatic creatures while decreasing waste and resource ingesting. For example, groundbreaking feeding systems driven by AI sense the precise starvation level of fish populations, thus averting overfeeding, which conventionally led to both economic loss and environmental dilapidation (Misimi et al., 2016). In addition, extrapolative analytics empower the prediction of illness jeopardies, permitting for defensive interferences rather than sensitive actions, which suggestively recovers animal well-being and biosecurity (Bostock et al., 2010). By way of international seafood mandate remains to increase, precision aquaculture suggests an achievable and longevity result that aligns with the increasing prominence of environmentally friendly accepted food manufacturing systems. Encounters persist, predominantly concerning the high charge of technology acceptance, data incorporation across platforms, and the need for expert people. Nevertheless, with continuing modernizations in inexpensive radar technologies, edge computing, and open-source AI models, the blockades to precision aquaculture are expected to moderate. Contemplating a partnership among marine biologists, data scientists, engineers, and representatives will be decisive in mainstreaming precision aquaculture worldwide. This archetype change not only enhances aptitudes, higher productivity, and higher-quality products but also indicates a critical step toward accountable aquaculture that equilibria productivity with ecological stewardship (Hegde et al., 2023). As AI endures to develop, its incorporation with aquaculture will unquestionably lead to more intelligent, stronger, and longer-lived aquatic food systems soon.

AI in sustainable and eco-friendly aquaculture practices

AI is serving a progressively revolutionary role in encouraging acceptable and environmentally friendly aquaculture applications. By way of international mandates for seafood upsurges, the longevity management of aquatic possessions converts supreme to avoid environmental dilapidation and guarantee long-term output. AI-determined systems, including ML algorithms and predictive analytics, are being used to augment feeding practices, thus decreasing feed leftover and minimizing nutrient contamination in water bodies (Cao et al., 2022). Smart radars and IoT-grounded monitoring networks empower the instantaneous valuation of water quality parameters, such as dissolved oxygen, pH, temperature, and ammonia levels, permitting ranchers to take appropriate remedial decisions that preserve ecosystem balance (Kim et al., 2022). AI models are also used for early illness discovery via image acknowledgment and pattern analysis, curtailing the requirement for chemical handling and antibiotics, which have harmful effects on the aquatic atmosphere. Furthermore, AI-aided decision-making tools provide site selection and carry capacity analysis, confirming that aquaculture connections do not surpass the environmental restrictions of their environs

(Shi, 2023). Progressive drone investigation and computer vision technologies permit habitat mapping and biodiversity monitoring in and around aquaculture zones, offering critical visions into preserving the veracity of wild environments.

In addition, AI-driven supply chain management apparatuses augment the traceability and documentation procedures for sustainably cultivated harvests, encouraging eco-labeling and customer conviction. These revolutions provocatively subsidize to attaining the goals of accountable aquaculture, as delineated by the United Nations' Sustainable Development Goals (SDGs), predominantly SDG 14 (Life Below Water) and SDG 12 (Responsible Consumption and Production). Still, the incorporation of AI technologies must be accompanied by ethical deliberations, transparent data ascendency, and capacity-building enterprises to confirm unbiased admittance and application across various aquaculture settings. As AI technologies endure to progress, they are predicted to play an increasingly central role in driving the aquaculture industry toward a greener and more longevity future (Fig. 14.1).

Integration of AI with other technologies

The incorporation of AI with evolving and prevailing technologies is expected to transform the future of aquaculture, foremost to more well-organized, sustainable, and groundbreaking invention systems. AI, when joined with the IoT, permits the instantaneous monitoring of water quality parameters, including temperature, dissolved oxygen, pH, and ammonia levels, through interrelated smart radars. This incorporation simplifies active farm management by forecasting latent hazards and augmenting feeding approaches, in that way minimizing left-over and improving fish well-being (Martínez-Porchas et al., 2023). Correspondingly, AI, coupled with remote detecting and geographic information systems (GIS), empowers the spatial investigation of aquaculture sites, thus enlightening site selection, disease investigation, and environmental influence estimations (Shi et al., 2026). Robotics and automation, when improved with AI competencies, are progressively being applied for errands, including net cleaning, fish classification, and autonomous nourishing, significantly decreasing labor costs and refining working efficiency (Kim, 2022). Besides, when combined with AI, blockchain technology improves traceability and transparency across the aquaculture supply chain, confirming food security and increasing

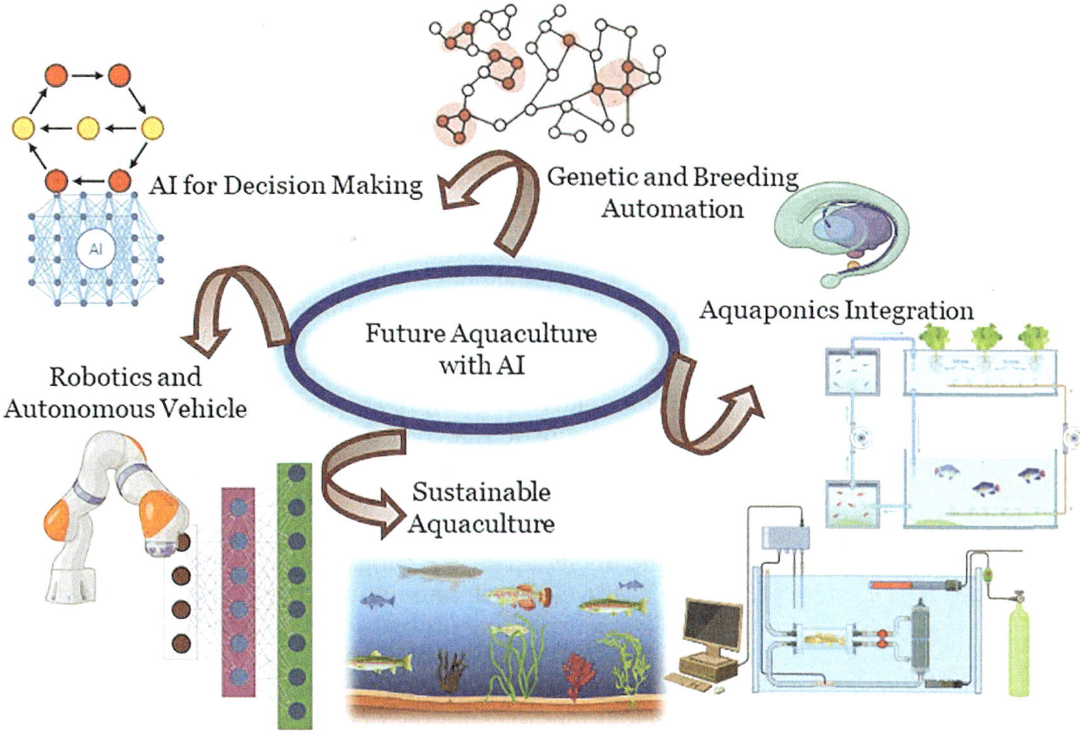

Figure 14.1 Future scope of aquaculture with Artificial Intelligence (AI) technologies.

customer assurance (Zhao et al., 2022). Current developments also highlight the interaction between AI and biotechnology, including utilizing AI algorithms to investigate genetic data for discerning breeding agendas, pointing to harvest disease-resistant and quickly growing aquaculture organisms (Sun et al., 2024). These interdisciplinary partnerships are not only improving output but are also vital in rectifying critical encounters associated with climate change, biosecurity, and resource optimization. Though efficacious incorporation demands disabling fences, including data standardization, infrastructure charges, and the need for upskilling the staff in digital knowledge. As aquaculture systems become progressively complex, the convergence of AI with IoT, blockchain, robotics, biotechnology, and big data analytics will form the backbone of next-generation aquaculture, lashing the sector toward resilience, longevity, and international food security.

Big data and predictive analytics in aquaculture

The incorporation of big data and prognostic analytics is quickly transmuting the aquaculture division, contributing groundbreaking resolutions to around of the manufacturing's utmost tenacious encounters. Contemporary aquaculture processes produce enormous amounts of data from varied foundations, such as water quality radars, feeding systems, well-being monitoring devices, weather stations, satellite imaging, and genomic data (Martins et al., 2013). The capability to leverage and investigate this "Big Data" offers a marvelous chance to augment manufacturing, improve longevity, and progress animal well-being. Prognostic analytics, driven by ML algorithms, predict illness epidemics, evaluate optimal feeding periods, forecast market demand, and anticipate environmental hazards with remarkable precision (Rao et al., 2021). Applying simultaneous data rivulets and ancient datasets, prognostic models permit ranchers to make well-versed, data-determined pronouncements, thus dipping losses, exploiting profits, and endorsing environmentally friendly practices. One capable application is the early discovery of pathogen epidemics, where prognostic models examine elusive vicissitudes in water restrictions and fish behavior to activate appropriate intrusions (Sun et al., 2020). Correspondingly, bg data analytics supports accurate aquaculture by adapting feeding schedules based on separate or group growth patterns, refining feed conversion ratios, and curtailing environmental leftover (Misimi et al., 2018). Cloud computing and IoT technologies additionally permit remote monitoring and control, permitting ranchers to accomplish large-scale processes with superior competence. As aquaculture expands to meet rising worldwide food demands, the upcoming will likely see even profounder incorporation of AI-determined prognostic analytics, enabling the modification toward ground-breaking, mechanized, and robust aquaculture systems. Yet, understanding the full potential of Big Data in aquaculture needs engaging encounters, including data standardization, privacy concerns, the requirement for interdisciplinary proficiency, and guaranteeing user-friendliness for small and medium-sized organizations. Inclusive, bg data and predictive analytics characterize an essential boundary in the development of aquaculture, flagging the way for a more industrious, sustainable, and intelligent industry landscape (FAO, 2020).

Challenges and limitations of AI implementation

The incorporation of AI into aquaculture offerings offers transformative opportunities, but until now, it has not been attended by numerous critical encounters and boundaries that to be prudently measured (Fig. 14.2). One of the leading challenges is the deficiency of high-quality, domain-specific datasets. AI models, particularly ML and DL algorithms, need massive quantities of explained data for training and authentication; though, in aquaculture, data collection is frequently fragmented, inconsistent, or inaccessible due to the diverse environmental conditions, species differences, and farming practices across provinces (Jensen et al., 2025). Moreover, the charge of employment remains a momentous barrier to acceptance. Backgrounding up AI systems requires substantial speculation in radars, imaging devices, computing infrastructure, and accomplished

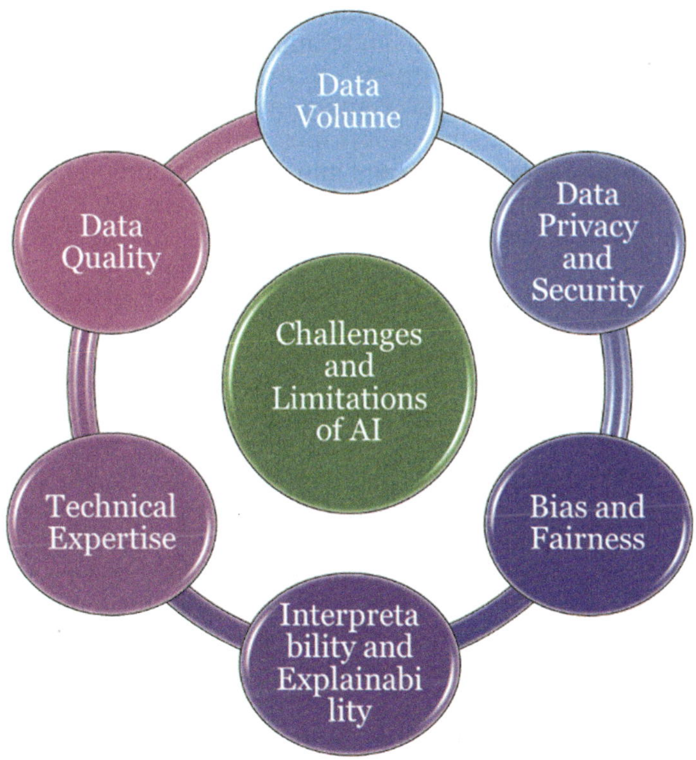

Figure 14.2 Various challenges and limitations of AI in aquaculture.

workers, which may be high-priced for small and medium-sized initiatives (Xie et al., 2022). Another constraint is the technical complication and preservation of AI systems. Aquaculture environments are dynamic and changeable, making it problematic for motionless AI models to familiarize without incessant retraining and recalibration. Additionally, there are moral and governing apprehensions, predominantly concerning the practice of independent executive systems in live animal management, which require severe strategies to guarantee animal well-being and environmental longevity. The interpretability of AI models, habitually denoted as the "black box" delinquent, also poses an encounter, as investors may vacillate in faith in AI-ambitious endorsements without transparent, understandable intellectual (Doshi-Velez & Kim, 2017). Finally, there is an acquaintance hole between aquaculture consultants and AI inventors. Many farm operatives may lack the technical proficiency to excellently deploy, troubleshoot, or augment AI explanations effectively, leading to underutilization or abandonment of exclusive systems. Addressing these encounters will necessitate a multidisciplinary determination connecting data standardization, reasonable technology explanations, education and training programs, regulatory outlines, and the expansion of explainable AI models personalized to the complex certainties of aquaculture. Upcoming research and strategy labors must emphasize these confines to empower a sustainable and impartial incorporation of AI into aquaculture practices.

Policy, ethics, and workforce transformation

The incorporation of AI into aquaculture is set to transform the industry; nevertheless, this transformation necessitates careful consideration of policy frameworks, ethical considerations, and labor force development. As AI-ambitious technologies, including ML, prognostic analytics, autonomous monitoring systems, and robotics, become progressively entrenched in aquaculture practices, policymakers must create clear guidelines to confirm the sustainable, reasonable, and transparent deployment of these inventions. Healthy governance constructions are desirable to safeguard data privacy, intellectual property rights, and environmental integrity while encouraging innovation and competition (FAO, 2022). Ethical encounters also arise, such as apprehensions around algorithmic bias, the potential domination of AI technologies by huge establishments, and the relegation of small-scale ranchers. Certifying that AI systems are industrialized and used dependably necessitates the founding of ethical strategies that highlight transparency, fairness, and inclusivity (Floridi et al., 2018). Furthermore, the distribution of AI in aquaculture will unavoidably reform the labor force. Outmoded characters absorbed in labor-intensive and experimental tasks will progressively give way to data-driven positions that necessitate proficiency in AI, data science, and system preservation. Labor force alteration creativities must arrange reskilling and upskilling programs to help prevailing workforces transition to new roles and avert extensive job dislocation (World Economic Forum, 2020). Educational institutions and industry stakeholders should collaborate to create interdisciplinary curricula that blend aquaculture sciences with digital and computational competencies. Furthermore, there is an unrelenting need for public-private partnerships to fund training programs, encourage digital literacy in pastoral and coastal communities, and guarantee that the aids of AI adoption are justifiably dispersed across the aquaculture value chain. As we move toward a future where AI plays a central role in aquaculture, aligning policy, ethical standards, and staff development policies will be crucial to building a robust, comprehensive, and maintainable industry.

Vision for the future: Smart, resilient aquaculture systems

The future of aquaculture mandates the incorporation of advanced technologies that generate groundbreaking, robust systems capable of rectifying the rising mandate for sustainable seafood production. As the worldwide population is expected to rise, the need for competent and maintainable food manufacturing approaches increases. Outmoded aquaculture airs abundant encounters, including environmental deprivation, disease epidemics, and resource restrictions. To confront these problems, the future of aquaculture will depend deeply on the positioning of AI, IoT, and advanced data analytics to transform the industry and confirm food security. An advanced aquaculture system exploits AI to screen and augment every feature of the manufacturing development, from water quality to fish health, feeding schedules, and environmental circumstances. IoT radars entrenched in aquaculture environments accumulate instantaneous data on parameters including temperature, salinity, oxygen levels, and fish movement. AI algorithms then examine this data to sense initial symptoms of disease, forecast growth patterns, and augment feed usage, thus reducing waste and refining feed conversion ratios. ML models also forecast environmental changes, including temperature instabilities or harmful algal blooms, allowing early interference and vindicating the risks related to these events. Moreover, the use of AI-determined automation in aquaculture systems is vital to refining operational efficacy. Automated feeding systems, robotic cleaners, and autonomous monitoring drones decrease labor charges, lessen human mistakes, and confirm that the system functions under optimum circumstances. The incorporation of AI and robotics will also improve the scalability of aquaculture ranches, permitting for more concentrated production without conciliatory the environment or animal well-being. A resilient aquaculture system is not only mechanically progressive but also capable of adapting to environmental, economic, and social stresses. As climate change endures to impact aquatic ecosystems, groundbreaking aquaculture systems must include climate-resilient technologies, such as closed-loop systems, sustainable energy solutions, and advanced water treatment technologies. These systems will reduce dependency on natural resources, mitigate the effects of pollution, and promote the restoration of degraded ecosystems. In addition, AI plays a vital role in emerging prognostic models for climate change impacts, permitting aquaculture processes to adjust proactively to changes in water quality and ecosystem dynamics. The incorporation of AI into aquaculture also unlocks new opportunities for the industry to align with broader longevity goals. Enhancing production effectiveness, minimizing resource usage, and decreasing the environmental footprint of farming procedures, intelligent systems subsidize to attaining the United Nations' Sustainable Development Goals (SDGs), particularly SDG 14 (Life Below Water) and SDG 2 (Zero Hunger). Furthermore, AI aids improve traceability and transparency in aquaculture supply chains, allowing customers to make knowledgeable adoptions and encouraging moral and supportable performs across the industry.

Conclusion

AI in aquaculture suggests transformative possibilities for the manufacturing, pushing it toward superior longevity, competence, and scalability. As we look to the future, AI's role in meticulousness in aquaculture will endure to grow, allowing instantaneous monitoring of water quality, fish well-being, and alimentation performs via the use of progressive radars and ML algorithms. These inventions will substitute augmented manufacturing systems, decreasing waste, minimizing disease epidemics, and enlightening resource management. Moreover, AI-ambitious predictive models will support decision-making processes, improving the capability to estimate trends in fish populations, climate impact, and market demands. With the continuing expansion of more robust AI tools and the adoption of automated systems, aquaculture has become more robust to environmental and economic challenges. As AI continues to progress, it will unquestionably redesign the aquaculture landscape, contributing new prospects for invention while addressing the growing worldwide demand for sustainable seafood production.

References

Bostock, J., McAndrew, B., Richards, R., Jauncey, K., Telfer, T., Lorenzen, K., ... Corner, R. (2010). Aquaculture: global status and trends. *Philosophical Transactions of the Royal Society B: Biological Sciences, 365*(1554), 2897–2912.

Cao, X., Xing, L., Majd, E., He, H., Gu, J., & Zhang, X. (2022). A systematic evaluation of supervised machine learning algorithms for cell phenotype classification using single-cell RNA sequencing data. *Frontiers in Genetics, 13*, 836798.

Dawson, T. P., et al. (2020). Opportunities and challenges for AI in aquaculture. *Aquaculture International, 28*(3), 1125–1143.

Doshi-Velez, F., & Kim, B. (2017, February 28). *Towards a rigorous science of interpretable machine learning* (arXiv preprint arXiv:1702.08608).

FAO. (2020). *The state of food and agriculture 2020: Overcoming water challenges in agriculture.* Rome: FAO. https://doi.org/10.4060/cb1447en.

FAO. (2022). *The state of food security and nutrition in the world 2022: Repurposing food and agricultural policies to make healthy diets more affordable.* Rome: FAO.

Floridi, L., Cowls, J., Beltrametti, M., Chatila, R., Chazerand, P., Dignum, V., Luetge, C., Madelin, R., Pagallo, U., Rossi, F., & Schafer, B. (2018). AI4People—An ethical framework for a good AI society: Opportunities, risks, principles, and recommendations. *Minds and Machines, 28*(4), 689–707.

Føre, M., Frank, K., Norton, T., Svendsen, E., Alfredsen, J. A., Dempster, T., ... Berckmans, D. (2018). Precision fish farming: a new framework to improve production in aquaculture. *Biosystems Engineering, 173*, 176–193.

Hegde, N., Kote, S., & Singh, A. (2023). Applications of artificial intelligence in aquaculture: a review. *Aquaculture Reports, 30*, Article 101598.

Houston, R. D., Bean, T. P., Macqueen, D. J., Gundappa, M. K., Jin, Y. H., & Johnston, I. A. (2020). Harnessing genomics to fast-track genetic improvement in aquaculture. *Nature Reviews Genetics, 21*(7), 389–409.

Jensen, D., Sønvisen, S. A., & Johnsen, J. P. (2025). A governability lesson for the Blue Economy: The collapse of mobile salmon aquaculture in the Norwegian coastal zone. *Maritime Studies, 24*(1), 2.

Kim, J., Kim, Y., Park, S. E., Kim, T. H., Kim, B. G., Kang, D. J., & Rho, T. (2022). Impact of aquaculture on distribution of dissolved organic matter in coastal

Jeju Island, Korea, based on absorption and fluorescence spectroscopy. *Environmental Science and Pollution Research, 29*(1), 553–563.

Kim, S. (2022). Working with robots: Human resource development considerations in human–robot interaction. *Human Resource Development Review, 21*(1), 48–74.

Li, Y., Zeng, Z., Lin, J., Wang, Q., & Zhou, L. (2020). Deep learning for aquaculture: status, challenges, and opportunities. *Reviews in Aquaculture, 12*(2), 924–941.

Martínez-Porchas, M., Vargas-Albores, F., Garibay-Valdez, E., & Martínez-Córdova, L. R. (2023). Advances and perspectives in integrated multi-trophic aquaculture. *Frontiers in Marine Science, 10*, 1197343.

Martins, P., Cleary, D.F., Pires, A.C., Rodrigues, A.M., Quintino, V., Calado, R., & Gomes, N.C. (2013). Molecular analysis of bacterial communities and detection of potential pathogens in a recirculating aquaculture system for *Scophthalmus maximus* and *Solea senegalensis*. *PLoS One, 8*(11), Article e80847.

Misimi, E., Ødegård, Å., & Eilertsen, A. (2020). Computer vision and machine learning in fish aquaculture. *Computers and Electronics in Agriculture, 170*, Article 105240.

Misimi, E., Olofsson, A., Eilertsen, A., Øye, E. R., & Mathiassen, J. R. (2018). Robotic handling of compliant food objects by robust learning from demonstration. In *2018 IEEE/RSJ international conference on Intelligent Robots and Systems (IROS) 2018 October 1* (pp. 6972–6979). IEEE.

Misimi, E., Østvik, S. O., & Mørkøre, T. (2016). Computer vision-based inspection and sorting of salmonids with emphasis on quality grading: a review. *Computers and Electronics in Agriculture, 123*, 129–139.

Neto, A. C., da Silva, J. S., Silva, A. A., & de Souza, R. M. (2021). Internet of things in aquaculture: a review on the current status and technological perspectives. *Aquacultural Engineering, 92*, Article 102146.

Qin, Y., Shi, C., Zhang, H., & Ma, D. (2020). Real-time monitoring for aquaculture based on computer vision technology: a review. *Aquaculture, 531*, Article 735923.

Rajesh, M., et al. (2021). Smart aquaculture monitoring using IoT and AI. *Aquacultural Engineering, 92*, Article 102182.

Rao, D. N., Rao, T. B., & Machiraju, P. V. (2021). Chemical and pathogen impacts on human health near aquaculture areas in West Godavari District of Andhra Pradesh, India. *Nature Environment and Pollution Technology, 20* (3), 1285–1293.

Rokhlin, S. I., et al. (2021). Feeding optimization in aquaculture with AI. *Frontiers in Artificial Intelligence, 4*, Article 654739.

Shi, B., Jin, X., Hu, Y., Jiang, J., & Sun, Y. (2026). Precision prediction of aquaculture water quality: A spatiotemporal model integrating optimized-LSTM and radial basis function neural networks. *PeerJ Computer Science, 12*, e3515.

Shi, C. (2023). Aquaculture environment regulation and system engineering. *Frontiers in Marine Science, 10*, 1236868.

Shi, W., Cao, J., Zhang, Q., Li, Y., & Xu, L. (2019). Edge computing: vision and challenges. *IEEE Internet of Things Journal, 3*(5), 637–646.

Sun, H., He, Z., Zhang, M., Yen, L., Cao, Y., Hu, Z., Peng, Y., & Lee, S. Y. (2020). Spatial variation of soil properties impacted by aquaculture effluent in a small-scale mangrove. *Marine Pollution Bulletin, 160*, 111511.

Sun, Y., Li, L., Meng, Z., Sun, H., Cao, R., Liu, R., Wang, S., & Liu, N. (2024). Rapid on-site differentiation of turbot from different culture modes using

miniaturized near infrared spectroscopy coupled with interpretable machine learning. *Microchemical Journal, 207,* 111981.
World Economic Forum. (2020). *The future of jobs report 2020.*
Xie, Y., Zhao, H., Li, J., Zhang, F., Wang, S., Yin, Z., & Shen, W. (2022). Phytoplankton biomass variation after cage aquaculture removal from the Daheiting Reservoir, China: Observations from satellite data. *Hydrobiologia, 849*(21), 4759–4775.
Xu, W., Yang, Y., Guo, Y., & Sun, W. (2023). Emerging technologies for smart aquaculture: Digital twins, blockchain, and AI synergies. *Journal of Marine Science and Engineering, 11*(1), 25.
Zhang, X., Liu, Y., & Zhang, L. (2022). Applications of artificial intelligence in aquaculture: a review. *Aquaculture, 548,* Article 737635.
Zhao, C., Sun, J., Gong, Y., Li, Z., & Zhou, P. (2022). Research on the blue carbon trading market system under blockchain technology. *Energies, 15*(9), 3134.
Zhao, X., Zhu, Q., & Huang, Z. (2021). Applications of deep learning in aquaculture: a review. *Aquaculture Reports, 19,* Article 100568.

Index

'*Note:* Page numbers followed by "f" indicate figures.'

9780443337048